To Ryan,

Service

By Michael L. Makfinsky

Cryptologic Technician Interpretive, First Class, (CTI1) – USN

With great admiration for all of the good work you perform to further Honorable Jamie Raskin's superb political drive!

Best,

Mike

ISBN: 978-0-578-21862-5

Printed in the USA
Signature Book Printing, www.sbpbooks.com

This work is dedicated to Victoria, my wife of intellect, stamina, valor, temperance and boundless love for our sons and grandsons, in appreciation for her having taken the "leap of faith" into what in Spanish is simply called "La Buena Vida Navy" – thank you, my Love.

To Jeanne - mother, your support was like water to baptism, kept everything REAL!

And to Ivan, Paul, and Daniel, we plied much world together, and I know that you will keep your sons curious and optimistic about tomorrow.

El Puerto de Santa Maria, Spain -2016-18

Cloverly, Maryland, USA - Mar 21, 2019

Contents

Foreword

Our most valuable legacy to new generations is the transfer of knowledge, and that's precisely the foundational purpose of this narrative. I wrote enthusiastically and researched methodically over the course of several months only because in my mind I'm telling this story to my sons and grandsons, who I want to thrill with amusing, if not artfully-crafted prose. These are my experiences as a Cold Warrior, and the narrative is intended to shed light on this important phase of our military's evolution.

If there is a lesson to be learned by my younger readers (and it's my sincere expectation that there is), let it be that life has many twists and turns in store for them, and that their intellect (*cogito, ergo sum*) and spirit, what the French call "*joie de vivre*", will govern the choices that they make in shaping their fate.

Beyond that basic intent, ***Service*** is written to convey the power of the individual in the context of a much larger socio-economic scheme and geo-political structure, and to juxtapose an individual serviceman's or servicewoman's personal efforts and sacrifices, as well as those of his or her family and friends, against the grand stage of world affairs. This is not to say that one plane of existence (i.e. the world of friends and family) is subordinated, or lesser than the vast maelstrom that is the world of international affairs - it is not, as we are all well aware that the fondest treasures we carry to our graves are more about the preciously exceptional emotional experiences shared with family and friends, rather than those impersonal triumphs or defeats that we might have harvested on the public stage.

This is an attempt to describe how the many countless lives that people have dedicated to *Service* make possible the outcomes that our elected leaders relentlessly pursue in this grandest of Republics - the United States of America. It is a narrative about the individual sailor's contribution to successful mission outcomes.

Yet another internal, more visceral and personal desire of mine is that this story may serve to counter the deleterious effects that actions by a notorious few, represented in part by the John Walker, the Chelsea Manning, and the Ed Snowden personas, have done to cast a shadow on what their many comrades in arms and colleagues in the public sector are so selflessly pursuing on a daily basis for the good of our nation:

Service

Preface

It's not about how you start.

It's not about where you end up.

It's about what you take with you.

It's about what you have given to your Country, and its citizens -children, women and men-, over the course of your finite time on Earth.

A Career of "Service"

The moral fabric of "Service" to your Country is lofty, altruistic, perhaps even utopian. Take a moment to grasp all of our Service mottos:

U.S. Navy = "Non sibi sed patriae" (Latin for "Not for self but for country")
U.S. Marine Corps = "Semper fidelis" (Latin for "Always faithful")
U.S. Army = "This we'll defend"
U.S. Air Force = "Aim high... Fly, fight, win"
U.S. Coast Guard = "Semper paratus" (Latin for "Always ready")

Now, if that doesn't make warm blood rush to your heart, and a sense of patriotic pride bring tears to your eyes, then you haven't thought about it long enough... "Not for self, but for Country" suggests that you go back and try again, and while you're at it, take some time with the U.S. Navy Ethos:

> We are the United States Navy, our Nation's sea power - ready guardians of peace, victorious in war.
>
> We are professional Sailors and Civilians - a diverse and agile force exemplifying the highest standards of service to our Nation, at home and abroad, at sea and ashore.
> Integrity is the foundation of our conduct; respect for others is fundamental to our character; decisive leadership is crucial to our success.

We are a team, disciplined and well-prepared, committed to mission accomplishment.

We do not waver in our dedication and accountability to our Shipmates and families.

We are patriots, forged by the Navy's core values of Honor, Courage and Commitment.

In times of war and peace, our actions reflect our proud heritage and Tradition.

We defend our Nation and prevail in the face of adversity with strength, determination, and dignity.

We are the United States Navy.

I. How You Start

There is such a thing as "juvenile enthusiasm", and science has proven that it's directly tied to raging hormones. History has it that Napoleon Bonaparte displayed "juvenile enthusiasm" when, after enlisting in the Army, and while attending the play "William Tell", he burst out yelling "Liberty, Liberty!" echoing William, the play's hero, after he pushes Geissler from the rock. Raging Napoleonic hormones? You be the judge.

My juvenile enthusiasm also led me to voluntary enlistment in the military. In retrospect this fateful event came to pass not only because I came from a military family and lived in a thoroughly military environment, it was also largely due to the fact that several of my 1968-1971 High School friends were grief-causing casualties in the Viet Nam war. The sadness that fellow classmates and I felt knowing that some of us had suffered a violent death while at war in a distant, foreboding foreign land tore a deep and painful scar in our psyche. Our thoughts were confused, defiant, leading to brash action when attempting to quench that inner fire.

Following a string of post-high school odd-jobs that I worked to cover for college tuition (ski shop salesman; musician in a rock band; gas station attendant), I made up my mind that I would beat the draft and avenge my friends' demise by enlisting in the U.S. Navy. Not even Country Joe & The Fish's chant, an anti-war protest song popularized at the 1969 Woodstock Music Festival, would hold me back:

> And it's one, two, three, WHAT ARE WE FIGHTIN' FOR?
> Don't ask me I DON"T GIVE A DAMN
> Next stop is VIET NAM.
> And it's five, six, seven, – open up them PEARLY GATES, well...
> There ain't no time to wonder why – WHOOPEE, WE'RE ALL GONNA DIE!

The lyrics convey the emotional funk that had hijacked my generation. It's a sad song for a sad generation of Americans who had lost all faith in the "system". The full measure of Country Joe's protest in part describes our Nation's predicament at that time:

"I Feel Like I'm Fixin' To Die Rag"

Well, come on all of you, big strong men,
Uncle Sam needs your help again.
He's got himself in a terrible jam
Way down yonder in Vietnam
So put down your books and pick up a gun,
We're gonna have a whole lotta fun.

Well, come on generals, let's move fast;
Your big chance has come at last.
Now you can go out and get those reds
'Cause the only good commie is the one that's dead
And you know that peace can only be won
When we've blown 'em all to kingdom come.

Come on Wall Street, don't be slow,
Why man, this is war au-go-go
There's plenty good money to be made
By supplying the Army with the tools of its trade,
But just hope and pray that if they drop the bomb,
They drop it on the Viet Cong.

Come on mothers throughout the land,
Pack your boys off to Vietnam.
Come on fathers, and don't hesitate
To send your sons off before it's too late.
And you can be the first ones in your block
To have your boy come home in a box.

And it's one, two, three,
What are we fighting for?
Don't ask me, I don't give a damn,
Next stop is Vietnam;
And it's five, six, seven,
Open up the pearly gates,
Well there ain't no time to wonder why,
Whoopee! we're all gonna die.

It was common practice for guys my age to beat the draft by enlisting in the service of one's own choice. As a matter of fact, several of my friends, neighbors of mine, had done just that, and told me that the Navy was the ticket to sure survival. Not only would you see the world by joining the Gulf of Tonkin Yacht Club, they said, but you would be assigned to ship's crew, or aviation, or logistics... all options that were a far cry from the Army and Marine Corps' infamous infantry purgatory. Others in my predicament would simply burn their draft card in protest, and head to Canada or elsewhere. Still others whose parents were high-up on the totem pole, would find a way out with deferrals, or cushy National Guard assignments to tide them over this raging, life-threatening military conflict. We all collectively, however, had vivid images of the gruesome My Lai massacre[1], and dreaded having to experience anything even remotely close to it.

---

[1] http://www.history.com/topics/vietnam-war/my-lai-massacre Retrieved Oct. 12, 2016 - In one of the most horrific incidents of violence against civilians during the Vietnam War, a company of American soldiers brutally killed the majority of the population of the South Vietnamese hamlet of My Lai in March 1968. Though exact numbers remain unconfirmed, it is believed that as many as 500 people including women, children and the elderly were killed in the My Lai Massacre. Higher-ranking U.S. Army officers managed to cover up the events of that day for a year before revelations by a soldier who had heard of the massacre sparked a wave of international outrage and led to a special investigation into the matter. In 1970, a U.S. Army board charged 14 officers of crimes related to the events at My Lai; only one was convicted. The brutality of the My Lai killings and the extent of the cover-up exacerbated growing antiwar sentiment on the home front in the

The Viet Nam war was a proxy war. The USA had no interest in conquering or dominating the lands and peoples that were emerging from French colonialism in Indochina. The widely accepted American view at the time was that the USSR (Union of Soviet Socialist Republics), through its COMINTERN[2] was supporting the Vietnamese uprising led by Ho Chi Minh, who intended to build a united communist Viet Nam. From there, our political leaders asserted, the USSR's COMINTERN would inevitably expand its influence throughout the entire Indochina peninsula (to include Laos and Cambodia), posing a serious threat to the USA and to our allies, first and foremost the propped-up South Vietnamese military junta. This was the grist of lectures that we got from our foreign affairs college professors. The "domino theory" had been developed following the end of the Second World War, and it postulated that what had happened in Eastern Europe, where the Soviet empire had virtually annexed entire nation-states from 1945 to 1960[3], would inevitably happen all around the world – *unless*...

United States and further divided the nation over the continuing American presence in Vietnam.

[2] http://www.collinsdictionary.com/dictionary/english/comintern - Short for Communist International: an international Communist organization founded by Lenin in Moscow in 1919 and dissolved in 1943; it degenerated under Stalin into an instrument of Soviet politics. Also called: Third International.

3 http://www.theatlantic.com/international/archive/2012/10/how-communism-took-over-eastern-europe-after-world-war-ii/263938/ Iron Curtain: The Crushing of Eastern Europe, 1944-56, explores the gutting of local institutions and the murders, terror campaigns, and tactical maneuvering that allowed Moscow to establish a system of control that would last for decades to come.

4 https://www.boundless.com/political-science/textbooks/boundless-political-science-textbook/foreign-policy-18/the-history-of-american-foreign-policy-110/the-cold-war-and-containment-586-4260/ - Containment was suggested by diplomat George Kennan who eagerly suggested the United States stifle communist influence in Eastern Europe and Asia. One of the ways to accomplish this was by establishing NATO so the Western European nations had a defense against communist influence. After Vietnam and détente, President Jimmy Carter

Unless the United States and its Allies exercised ar campaign of containment[4] to keep the Soviet bear at bay be Curtain, whereby the ideological cancer that was *Commun* neutralized, perhaps even defeated.

This geo-political threat had been cleverly designed to engulf the capitalist democracies of the world, and our leadership asserted that Communism had to be stopped in Viet Nam – that was our Rubicon for committing America's lifeblood in a battle for ideological supremacy. We the American youth, were immersed in (and co-opted into) a protracted Cold War against a "Communist" dominance of planet Earth[5]. When you put that in perspective -nuclear weapons perspective, that is, whereby

---

focused less on containment and more on fighting the Cold War by promoting human rights in hot spot countries.

[5] Other Western allies of ours were also immersed in the Cold War. For an Australian perspective released in July 1988, and retrieved on Oct. 12, 2016 see http://www.ausairpower.net/TE-Sov-ASuW.html "the USSR has poured considerable resources into developing a capability to interdict Western SLOCs; the rapid growth in this capability over the last decade is however quite alarming. The Soviets have followed a three pronged strategy of acquiring exceptionally well placed forward bases and of modernizing their air and submarine maritime strike force. Cam Ranh Bay was a valuable prize, offering the ability to cut all lanes to and from Singapore, Taiwan and the Philippines, while forcing much traffic to Japan to take a distinctly Eastern approach. Soviet access to Ethiopia and Aden has similarly provided good coverage to the Indian Ocean approaches to the Suez Canal and Straits of Hormuz. Angola offers only limited coverage of the Cape of Good Hope routes but the potential of South Africa as a target for subversion and eventual domination has worrying implications. The Soviet thrust into Chile during the early seventies is less than coincidental in this context. The Reagan Administration's concern over Central America is not unreasonable given the threat posed to the Panama Canal. This situation is a direct result of a systematic effort aimed at providing the capability to destroy Allied economic strength through interdiction of its SLOCs; the former CinC of the Voenno Morskii Flot (V-MF = Soviet Navy) Admiral Gorshkov has clearly stated that the interdiction of merchant shipping and its associated escorts is the most important constituent part of the efforts of a fleet aimed at undermining the military/economic potential of the enemy". The document goes on to describe the V-MF power projection capabilities.

ꞁoth the USA and the USSR had enough nuclear weapons to trigger humanity's doomsday Armageddon- you might conclude, like I did, that military service was the only logical survival mechanism for you, for your family, and for your future progeny.

Of course, there was also a social and cultural context that framed my response to this global ideological confrontation. Hollywood and the media were doing a great job of conveying the kind of spy-vs.-spy thrillers that would capture the imagination of my entire late baby-boomer generation. Sean Connery as James Bond / Secret Agent 007 embodied a brand of cool heroism that American machos (and no doubt others world-wide) aspired to. Our television sets were chock-a-block with espionage-themed series – Man from UNCLE; Mission Impossible; Invisible Man; The Avengers; I Spy; Secret Agent Man; etc. Even sitcoms such as Maxwell Smart's "Get Smart" (one of Mel Brooks' hit shows), kept us focused on the onerous predicament that the world was in, while laughing out loud at Agent 86's (Maxwell Smart's) bumbling nature and demands to do things by-the-book. In fact, our media was so saturated with this type of thriller fiction that, for a weary adult population, satires became just as popular, and movies such as "Our Man Flint", and "In Like Flint" emerged to put the feminist movement into a battle of the sexes juxtaposition with the all-consuming Capitalism vs. Communism race for hegemony. Other films, such as "Dr. Strangelove", and "The Mouse that Roared" (with stellar performances by Peter Sellers), became box-office blockbusters by parodying the USA-USSR imperial rivalry. It's good that we could laugh at ourselves. That said, over 280,000 Allied troops died in the Viet Nam conflict that was waged during our youth. The Vietnamese deaths were almost twice that number. There's no denying that the war in Viet Nam was a moral catastrophe – just ask John Kerry.

I remember what my Dad, (a WW-II veteran, and a Korean War veteran), told me when I was nine years old: that war brought out the *worst* and the *best* in people. He also believed that wars were a necessary evil in society, since in his view they served to purge the population, which otherwise would grow out of control. Seen in that light, wars were humanity's version of natural selection. I still haven't bought into that reasoning, although I do believe in the concept of a "just" war (*jus bellum iustum*) as justification for certain wars, such as the Allies' response to Nazi warfare for conquest and subjugation.

Was the Viet Nam war a just war? The Just War theory has it that war can be justified if and when it is the only option available to avoid a greater calamity, such as the enslavement of a majority of the human race by an armed, despotic minority. Under those circumstances, the righteous majority would exercise just warfare to avoid tyrannical repression and exploitation. World War Two was by all measures a just war to ensure that Hitler and his Nazis could not impose their vision of a totalitarian global hegemony – the Third Reich.

St. Augustine himself originated the "Just War" phrase in his work "The City of God":

"But, say they, the wise man will wage just wars. As if he would not all the rather lament the necessity of just wars, if he remembers that he is a man; for if they were not just he would not wage them, and would therefore be delivered from all wars."

The historical baggage of the Viet Nam war, however, points first to French colonialism for economic gains, followed by a popular insurrection led by Ho Chi Minh, who in the name of a majority of Vietnamese people claimed to assert the Just War principle of Self Defense against French colonialism. Ho's quote: "It was patriotism, not communism that inspired me" fueled the Viet Minh toward the eventual unification of North and South Viet Nam in April 1975.

But was North Viet Nam true to the tenets of a just war? The answer to this question, an unequivocal NO, is provided by the eye-witness account of one of the survivors of the 1968 Hue Massacre: Alje Vennema, a Dutch-Canadian doctor who lived in Hue and witnessed the 1968 battle and the massacre (Ho Chi Minh was by then on his deathbed), and wrote *The Viet Cong Massacre at Hue* in 1976[6]. In his book he describes the many horrific atrocities committed indiscriminately against the local population by the Viet Cong, atrocities such as the execution of a 48-year-old street vendor, Mrs. Nguyen Thi Lao, who was "arrested on the main street. Her body was found at the school. Her arms had been bound and a rag stuffed into her mouth; there were no wounds to the body. She was probably buried alive". A 44-year-old bricklayer, Mr. Nguyen Ty, was "seized on February 2, 1968...His body was found on March 1st; his hands were tied, and he had a bullet wound through his neck which had come out through the mouth. At Ap Dong Gi Tay "110 bodies were uncovered; again most had their hands tied and rags stuffed in their mouth. All of them were men, among them fifteen students, several military men, and civil servants, young and old". "Sometimes a whole family was eliminated, as was the case with the merchant, Mr. Nam Long, who together with his wife and five children was shot at home. Mr. Phan Van Tuong, a laborer at the province headquarters, suffered a similar fate by being shot outside his house with four of his children". Vennema listed 27 graves with a total of 2397 bodies, most of which had been executed.

We Americans, perhaps blind-sighted by our leadership's Communism vs. Democratic Capitalism mantra had embarked on a desperate attempt to prop up a weak South Viet Nam government. Its weakness stemmed from corruption, and the South Vietnamese regime proved unable to match the resolve and commitment of the Viet Minh and the Viet Cong. While South Viet Nam and the USA also claimed to be enforcing the principle of Self Defense against North Vietnamese aggression, the 1968 Tet Offensive in which North Viet Nam attempted to overthrow the South Vietnamese government, succeeded in proving to the world that North Vietnamese and Viet Cong forces could withstand

---

[6] Vennema, Alje (1976). *The Viet Cong Massacre at Hue*. Vantage Press. ISBN 978-0533019243

the enormous might of the U.S. military. To quote encyclopedic resources with purported objectivity:

"The Tet Offensive persuaded a large segment of the U.S. population that its government's claims of progress toward winning the war were illusory despite many years of massive U.S. military aid to South Vietnam.

Gradual withdrawal of U.S. ground forces began as part of "Vietnamization", which aimed to end American involvement in the war while transferring the task of fighting the Communists to the South Vietnamese themselves. Despite the Paris Peace Accord, which was signed by all parties in January 1973, the fighting continued. In the U.S. and the Western world, a large anti-Vietnam War movement developed as part of a larger counterculture. The war changed the dynamics between the Eastern and Western Blocs, and altered North-South relations.

Direct U.S. military involvement ended on 15 August 1973. The capture of Saigon by the North Vietnamese Army in April 1975 marked the end of the war, and North and South Vietnam were reunified the following year. The war exacted a huge human cost in terms of fatalities. Estimates of the number of Vietnamese soldiers and civilians killed vary from 800,000 to 3.1 million. Some 200,000–300,000 Cambodians, 20,000–200,000 Laotians, and 58,220 U.S. service members also died in the conflict, with a further 1,626 missing in action."

Conscription or Enlistment

The ice-breaker question at Boot Camp was "Were you drafted, or did you volunteer?" Because the Recruit Training Center (RTC) in San Diego, California was Navy, one could guess the likely answer, since for the most part draftees usually went to either the Army or the Marine Corps. That said, when asked about his draft card number, one of my boot camp acquaintances confessed: "My dad knew the Judge, so when the marijuana charge stuck, the Judge asked me which service I wanted... and here I am." Rather than get a chunk of his life trampled on in a penitentiary, this opportunistic patriot chose a two-by-six hitch[7] in Uncle Sam's Navy.

My enlistment process was not typical. Living, working, and studying overseas in Madrid, Spain, I had opted to connect with the Naval Attaché at the US Embassy (through my dad, a retired naval aviator, with connections from being active in the local chapter of the Navy League), who managed to have a Yeoman from the Office of Defense Cooperation hammer out an enlistment contract that would cart me off to the nearest US Naval installation in Rota Spain. From Rota I would catch a hop on a military chartered aircraft to eventually check-in at the RTC in San Diego, California[8]. The trip from the San Diego airport to the RTC took a mere 10 minutes, and the bus was packed with a 50/50 mix of long-hairs (yup, 70's vintage hippies brought up on Doors, Jimi Hendrix, Cream, Rolling Stones, Led Zeppelin, Janis Joplin, etc.) small town country boys making good on a promise to mom and dad or to their high school sweetheart. Out of a bus-load of about 48 enlistees, maybe only a dozen of us sported hair that passed Navy grooming standards. The RTC barbers were going to have a field day on us, and fill a lot of fluff bags with recruit-manes.

Wake-UP Call

Twenty-gallon galvanized steel trash cans can serve many purposes. In this case, at 4:30 am Pacific Standard Time on April 4th 1974, the lids were used as cymbals, and the drums were hurled to crash into each other across the cavernous open space in the 74-man open bay barracks of what would soon become "Charlie Company".

---

[7] Two-by-Six Enlistment Contract: mandates that the first two years be on active duty, with the remaining six years in the Navy Reserve. Once the initial two years of active duty are completed, the service member has the option of extending or re-enlisting on active duty, possibly reducing the obligation to six years of active duty service.

[8] For a history of the RTC San Diego CA, Retrieved on Oct 12, 2016 http://www.quarterdeck.org/AreaBases/Recruit%20Training/RTC%20command_history.htm

"REVEILLE, REVEILLE – WAKE UP YOU MAGGOTS! RISE AND SHINE, BEAUTY QUEENS... YOUR BEAUTY SLEEP IS OVER!"

Even though the clangs, hollers, and commotion came from uniformed men acting as prison guards, we the uncut, unshaven recruits, though startled and bewildered, held back, shouting out defensive expletives in response:

HEY, MOTHERFU..ER – F..K. YOU AND THE HORSE YOU RODE IN ON! SHUT THE F..K UP AND LET A DUDE CATCH SOME ZEES!

Non-compliance was futile, however, and soon all scraggly Joes (the Janes, much fewer in number, were remanded to an entirely separate grid on the RTC map) were marching out of the barracks to the induction center (now called the in-processing center), run by the Quartermaster, where we would all sign for uniforms, supplies, and toiletries in lieu of pay. Our civilian "mufti" went into an individual bin (to be forwarded to someone by mail – no "civvies" were authorized until after graduation). Our garb was instantly replaced with two-tone blue RTC recruit uniforms. The assembly line had it that once your "skivvies", t-shirt, trousers, belt, watch cap, shoes and socks were on, you stuffed the rest of the "Government Issue" into a freshly stenciled Sea-bag, and you then got in line to have a barber haul down that freedom flag that had served as a critical part of your '70s ID.

Navy regulations allowed for no less than ¼ inch crew cuts, but in defiance, a set of twins insisted they would not leave until their heads looked like polishable billiard balls. Since they were both about 6'2", and were obviously varsity wrestling jocks, they got what they wanted, totally oblivious to the unintended consequences, however. In boot camp it's easy to spot recruits that don't fit the standard mold: too tall, too short, too thin, too fat, and yes, you guessed it - too bald! These outliers were to get denigrating nicknames and experience the kind of heckling and harassment that inmates put up with in the pen. Needless to say, there were only two bald guys in Charlie Company.

"YO, Bobsies – Are you really that bald, or are your necks just blowing bubbles?!"

Enter the Recruit Petty Officer (RPO), selected by the Company's Chief Petty Officer (CPO, a real active duty non-commissioned officer) to help with discipline, and to serve as the CPO's eyes and ears on personal matters. The RPO, in this case a black African-American guy from somewhere near Chicago, had the temper of a pit bulldog, and rugged looks and ripples to go with the temper. The twins just stared back, which was enough to put a stop to anyone in the barracks even considering a taunt.

"Boot camp doesn't last forever" came the reply - the bobsies had him. It was a big Navy we would soon join, and as sailors unconstrained, accounts could and would be settled.

Grinders and Socials

There's an official schedule that describes what recruits learn during the eight weeks of recruit training[9], with watch standing basics, care and maintenance of issued gear during the initial "processing week". Week one concentrates on physical conditioning and Navy core values; week two includes chain-of-command, traditions and military courtesies; week three deals with laws of armed conflict, basic seamanship, shipboard communication, and Navy ship and aircraft identification. During week three recruits also take their first physical training test, performing as many sit-ups and push-ups as they can in two minutes and a mile and a half timed run. Week four includes weapons training; week five is called hell week because it's crammed with military drills and facility and equipment inspections; week six has damage control and firefighting instruction, to include use of the self-contained breathing

---

[9] For additional information on Navy boot camp experience see http://www.bootcamp.navy.mil/about.html Retrieved Oct. 12, 2016.

apparatus (SCBA), and exposure to the Confidence Chamber (a tear gas chamber) where recruits learn just how effective the SCBAs are.

The RTC instructor would ensure that only about a dozen sailors entered the Confidence Chamber at a time. The rest would stand-by wondering why the heck those that exited from the other hatch on the far end, beyond our field of view, were gagging, sobbing, and in some cases wheezing like wind bags.

"NEXT GROUP, don your SCBAs, and make darn sure that you have a tight seal around your face... no hair in the way of the rubber mask!" An unnecessary precaution, since we male recruits at this point had absolutely no worries about hair. Upon graduation we would shred the shackles of boot camp prison and enter into the "Zumwalt Navy", where grooming standards would shift considerably.[10]

The instructor would then open the hatch to this outdoors steel chamber, which simulated a shipboard compartment.

"Next group: IN! Enter the compartment smartly... zipper formation; backs of your hands sensing the bulkheads to determine temperature. Avoid the heat. Use your SCBAs as instructed".

---

[10] Admiral Elmo Zumwalt assumed duties as CNO and was promoted to full Admiral on July 1, 1970, and quickly began a series of moves intended to reduce racism and sexism in the Navy. These were disseminated in Navy-wide communications known as "Z-grams". There were seventy "Z-grams" promulgated during the period July 1, 1970, through 21 January 1971. Seventy proclamations were issued in a two hundred five-day period or a Z-gram once every 2.93 days, on average. Massive change so rapidly had the result of upending everything that tradition, daily shipboard routine, Navy discipline, grooming standards, and personal relationships, developed over almost two hundred years, was based upon. Z-gram 70 (21 January 1971): clarified grooming standards and working uniform regulations addressed by Z-gram 57 to reflect contemporary hair styles and allow wearing working uniforms while commuting between the base and off-base housing. Source: Garland Davis, "Tales of an Asian Sailor" https://garlanddavis.net/2016/04/15/admiral-zumwalt-and-the-z-grams/ Retrieved on Oct 12, 2016.

The compartment was dark, eerily lit by only a couple watertight marine lamps on the ceiling, emitting red light – it reminded me of a photography dark room. All twelve recruits were inside now, and the instructor who accompanied us had battened down the hatch behind him. His muffled voice came from behind his SCBA:

"See the emergency exit in the back of the compartment? Yeah? Well, when I give the order, I want all of you to assemble near that exit – understood?"

We recruits gave a knee-jerk thumbs up.

"Remove your SCBAs! Do it, and DO IT NOW!"

The instructor kept his on, of course, and we recruits, like babes in the woods, experienced tear gas, front and center.

The thing about tear gas is that it first slams your nostrils and throat, shutting them down so that you have to really make an effort to suck in a new lungful of air. But when that new lungful comes, you wish you hadn't inhaled because of the sting and burn it causes... sure, the eyes reflexively jam shut and burst into tears, and some recruits got kicked in the stomach, barfing into the SCBAs they were holding.

"Alright – that simulated chemical attack proves just how important your SCBA is for survival. This could have been mustard gas, or worse: SARIN!" He barked through the mask. He then took it off, for effect, maybe, and holding back tears shouted: "Now, who can tough it out and go for three more minutes in this Swedish sauna?" A couple bone-heads mumbled, gasping, that they could take more punishment.

"You two stay here... the rest of you get your masks back on!" Those of us who could did, and those who had barfed into their mask joined the Swedish sauna clowns. From the protective comfort of our SCBAs we could see their eyes puffing up, and their coughing intensify. The three minutes passed – an experience we would draw from for shipboard readiness.

“Assemble at the exit hatch!” The shaken dozen became a loose formation near the hatch, and the instructor opened it to let us quickly abandon this sour prison. The effect of being back out in the open surrounded by fresh air was cleansing. Our lungs were reborn.

In boot camp parlance, the Confidence Chamber is a “grinder”, an experience designed to instruct and build character. There are many other such “grinders” to chalk up, and one of them left a lasting impression on me: the “Midnight Grinder”. How I got on the “Midnight Grinder” roster is simple: I challenged the RPO. Not to fisticuffs or anything violent, I merely stated out loud, in formation, that he didn’t know what the heck he was talking about, a.k.a. the *military brat* in me tossing a bullshit red-flag out onto the field. When I was informed the next day that I was scheduled for additional duty – Midnight Grinder, I asked about it, but got no explanation, other than “You’ll see – it’s a whole lotta FUN!”

Positional leadership can’t be bucked in boot camp, and here’s how this kind of an infraction is dealt with:

When your name appears on the Midnight Grinder list it’s always because your Company CPO put it there, usually to “correct” disciplinary problems, such as failure to be on time for watch, challenging authority, or minor code infractions (major infractions are dealt with legally, possibly resulting in a Big Chicken Dinner – a euphemism for Bad Conduct Discharge, which is a label that stays with an ex-GI forever), etc.. At 23:30 on a brisk San Diego Bay night, I was roused from my “rack”, and told to get my Physical Training (PT) gear on, and to assemble in front of the “Quarterdeck” (official place of business of a Navy unit – in the case of the RTC, the entrance to the story of open-bay barracks) within TWO minutes. My PT gear consisted of a T-shirt, shorts, socks and sneakers, and, since it was nighttime, a watch-cap. I was at the QD on time. A roving “recruit formation” led by an RPO (not mine) came by within minutes.

"Join the FORMATION - **Hooah**!" barked the RPO, and three of us miscreants from Charlie Co. fell-in toward the rear of the formation, running at half pace. This process was repeated as the formation approached the various Company Quarterdecks on the way to the Grinder.

Turns out that the "Grinder" was a large paved open area with a tall flagpole out front. A huge star spangled banner – National Ensign to us, waved proudly atop the flagpole in the cold, humid San Diego breeze. In our short time there, none of us recruits had been allowed to venture out onto this part of the RTC, and we later found out that it was in fact the parade ground where our proudest RTC moment – the graduation Pass in Review[11], would take place.

On a platform beneath the Flag stood a buff guy in black PT gear. He addressed the formation over a PA system: "Formation... (the loud-speakers reverberated, reminding me of those Hitler Nuremberg rallies I'd seen in my High School world history class)... Ah-ten-HUT!" We recruits popped as tall as we could, thumbs flush with the seams on our shorts. A cadre of about twenty gruff Petty Officers encircled the formation, walking slowly like pumas ready to pounce on their hapless prey. "Is it too cold for anyone? If so, raise your hand!" barked the emcee. Several recruits had not read the proverbial memo: "In Boot Camp don't ever volunteer, and don't ever, EVER complain", and their naïve hands shot up. With the stealth of a night prowler, the Petty Officers encircling the formation (we later found out that they were drill instructors at the nearby Basic Underwater Demolition School (BUDS) in Coronado – the birthplace of the elite Navy Sea-Air-Land sailors, better known as SEAL Special Forces Unit), zeroed in on the "complainers", and struck their wool watch-caps off their skulls and onto the grinder. "There – that's MUCH better" grinned the emcee. "And now, Formation,

---

[11]http://navyformoms.com/custom/media/downloads/N4M_PIR_infographic_Printing.pdf. Retrieved Oct. 12, 2016. Describes the importance of the Pass in Review ceremony.

Jumping Jacks in place – FOREVER. **GO** – I want to hear the count LOUD AND CLEAR!" That went on for about 20 minutes, with the roving PO's ensuring that all rebels gave a snappy clap of the hands at the top of each jumping jack.

"On your bellies, WORMS! Push-up position. NOW PUSH-UPS FOREVER. **GO!** – I want to hear the count **LOUD AND CLEAR!**" Most of us could easily give about 40, but from that point on, the arms tightened up. The POs knew what to do… as they circled their prey, one would place a boot under the slovenly recruit's gut, while another would place a boot over the slovenly recruit's shoulders, and they would renew the count for their hapless victim: "One WORM, Two WORM, Three WORM!", and so on as the boots sometimes worked in favor of a successfully easy push-up, and other times completely against the arms' extension or the arms' flexion. When a recruit complained about exhaustion, both boots would get placed on his back: "WORM, get those arms MOVING, WORM!" "You're in the U.S. Navy NOW, WORM!" Slack-off now and we'll make sure you start ALL OVER AGAIN, WORM!"

For some reason, having a couple boots shoving you into the ground gets your goat every time. Angry growls and fierce yells were heard throughout the entire formation, as recruits (rebel recruits, at that) strained their over-exerted arms to the last sinew to complete just one more "burdened" push-up. In their heads these recruits had a parallel push-up count going: "F..K YOU – One; F..K YOU – Two; F..K YOU – Three… Even so, we were being pushed well beyond any point of exhaustion we had known in our lives.

When the night-time mist turned to cold drizzle, the emcee announced: "How LUCKY for you, WORMS. Gotta shut off the mike. RPO – TAKE CHARGE OF THE FORMATION! FORMATION – DISMISSED!" As quickly as we had been put to test, we were now being released back to the comfort of our Company barracks.

Following a quick run through the rain, we reported back to our respective QDs, and in my case the Sentry made the following entry in Charlie Company's Log-book:

"01:07 April 22, 1974 – Recruits Baird, Makfinsky, and Salazar report back from Midnight Grinder duty."

My April 22 05:00 Reveille would be just another rude awakening.

The lighter side of boot camp was the social gatherings near the barracks, in an area set aside as a smoking haven called "Smokers". Recruits would assemble loosely to "smoke 'em if you got 'em"; to "shoot the breeze"; to dispense "scuttle-butt"; etc. What was not permitted was "sky-larking" or "lolly-gagging" (you might want to waste your time looking up these boot camp slang terms).

There was another kind of "Smokers" that assembled on Friday evenings, after dinner, where recruits from the entire RTC also puffed on coffin nails to their hearts' content while watching other recruits go at it in a boxing ring. One of my fellow recruits in Charlie Company was a Golden Gloves boxer, so he was only allowed into the ring with other Golden Gloves, and sadly for him, there were never any takers. The rest of us pukes were encouraged to put on the gloves and duke it out any way we could, the "winner" guaranteed easy duty on Sunday after whatever religious service one might attend (and believe you me, we all did attend a Service, or otherwise cleaned the barracks).

Not shy when challenged, I once stepped into the ring, and gloves on, threw some jabs around at another wanna-be Mohammed Ali. The first three-minute round was ok, the second round was grueling, and at the start of the third round I took an upper-cut to the jaw that instantly turned off my vision. My jabs flailed until I landed some body punches, but the ref, seeing that I was for all intents and purposes blind, stopped the fight. Ref then hoisted the other guy's arm high, proclaiming a winner as I protested my lack of vision. A wasted effort it was, so my only consolation that Sunday came in attending the Jewish Service (I know, wrong day for Jewish observance, but that's how it was done in boot camp) with two other gentile compadres. Don't read anything into this... we went not only out of curiosity, but mainly to egg-on the Rabbi into pouring us several shots of Mogen David sweet wine for us *"Shegetz's"*. Seeing what we were up to, the Rabbi chuckled, and poured generously.

That was the only time I ever got a buzz in boot camp.

Regular "Smokers" alongside the barracks were much more low-key. At these gatherings one could hear about things like how a particular recruit's high-voltage lawyer dad was going to get his enlistment contract cancelled for breach of contract; how a particular recruit's mother sent him jars of Mexican miniature pickled jalapeno peppers; how a particular recruit got a "Dear Johnny" letter from his high school sweetheart; what the guys on Kitchen Patrol (KP) duty would add to the mashed potatoes to suppress sexual libido ("saltpeter" – a possible class action suit?). The list of amusing, if not fascinating topics goes on, but you'll get a kick out of the Mexican jalapeno (Spanish - *jalapeño*) story.

The care package with said jars (maybe four 32 oz. jars) had gone on display at "Smokers" – "Hey guys, my Mom sent me four jars of my favorite jalapenos!" Jack advertised to all within hearing range. Jack was a construction worker who had signed up for a stint as a SeaBee (*which comes from the acronym CB, for Construction Battalion, so Navy construction sailors are known as SeaBees*). "Man these peppers are GOOOD!" Jack said gleefully, brandishing a jar high above his head.

No shortage of hucksters in boot camp, believe me, and Lenny - the guy who enlisted in lieu of a MaryJane sentencing- replied: "If they're so damn good, I'll give you $10 to eat one of those jars, whole, in one sitting! BETCHA CAN'T!"

Now we had a session going – this is what constituted **great** entertainment at the Smokers. Other recruits chimed in (including me): "Yeah, Jack – I'll also put down $10 to see you pepper that big gut of yours." Nine of us bet him $10, so he suddenly stood to rake-in some good liberty dough (these were 1974 dollars!)

To Jack, who had obviously never done this, but who suddenly realized that he could enact a double feature: quench his obsessive hunger for good jalapenos while earning $90, the course of action was clear: "You're ON, MotherFu..ers!"

So Jack popped the lid on the 32 oz. jar, and proceeded to drain the pickling juice into a storm gutter. Lenny pointed out that Jack's failure to down the jalapenos in one sitting would result in a pay-out of nine times ten dollars.

"No SWEAT, dude – I got it! Now someone round up the $90 so I can get started."

As soon as Lenny showed Jack the money, down the hatch went the miniature jalapenos – by the mouthful, and Jack chugged out of the jar as if it contained a cool, frothy beer.

When Jack got about two thirds of the way into the jar, he sputtered: "Water. I need WATER!" You guessed it; Lenny just stared back at Jack, smiling like a Cheshire cat, in anticipation of his pay-out. One of Jack's buddies, who had also bet against him, was quick to shove a cup of water toward him. Jack gulped the water down. His eyes were now bloodshot, tears streaming down his face onto his recruit work shirt, staining it with big blotches of dark blue. "DAMN THESE ARE GOOOOD!" he belted out nonetheless as he proceeded to chug another mouthful of demon peppers. The going was painfully slow as he dealt with that last one third of peppers.

Finally, with the bravado of a matador, Jack held the jar at arms' length directly in front of his face, and defied his spectators: "Throw in another $10 for the grand finale, suckers!"

"Hell naw – a bet's a bet. You leave that in the jar, and you'll pay up, guaran-fuk..g-teed!" Lenny said.

The jar went bottoms up. Jack was sweating profusely, undergoing a transformation that only he could comprehend while the rest of us jeered him on, willing to give up a Hamilton note for what was so far the best entertainment that money could buy in this place and point in time. When the last pepper disappeared from the jar into Jack's jaws, he lunged at Lenny, grabbed the $90 and tore out of The Smoker like a bat out of hell, heading right back to the barracks.

We never found out whether he chewed and ate the final mouthful (Jack swore he did), but we did hear his moaning through the night, when Jack experienced jalapeno burn on its way out, as he got "tore a new asshole".

Interviewed by NCIS

The acronym stands for Naval Criminal Investigative Service, and, with the consent of Commanding Officers, NCIS gets investigative powers over Navy personnel. By the first week of boot camp, three of us "worms" had befriended, and almost on a daily basis shared views on how to get to the end of the tunnel without punching someone's lights out. All we had to do was to last the first two weeks on "Shit Island" – a cluster of barracks and a chow hall well isolated from the rest of the RTC by a moat of seawater that was 20'-30' deep in some parts, and connected to the RTC only via "Liberty Bridge", which had a look-out on duty 24/7.

Both Dan and Brad were Californians, kind of laid back as compared to the rest of the company. Dan's family had urged him to get off his lazy high school grad butt, and see the world. Since his uncle had been in the Marines, Dan talked to a recruiter and decided on going Navy. After graduation, Dan would invite me to stay at his family's place in Ventura, as a convenient "real world" breather on my way to the Defense Language Institute at the Monterey Presidio.

Brad, on the other hand, hailed from San Fran, and his dad, an attorney, had become disgusted with his wanton ways of drugs, love-in roomies, a.k.a. sex and rock&roll, so he gave him an ultimatum: If he wanted to eventually go to law school and join the firm, he had to first do a hitch in the military to work out the kinks in his despicable (to mom and dad) lifestyle. Brad said that he finally literally saw the writing on the wall while tripping on acid, and signed a batch of unintelligible enlistment papers at some San Fran Navy Recruiting Station. He says he never read the papers, but was sure that there was nothing in the contract about getting treated like a mangy dog. Dan and I agreed, but tried to persuade him that the best way to get through this was to grin and bear it. Two weeks was nothing in the grand scheme of things.

Somehow that was no consolation to Brad, who became more withdrawn and taciturn as the days went on. Daily squabbles with instructors and chain of command that to Dan and me just went with the turf, were stinging transgressions to Brad, and he had to find a way out, come hell or high water.

At reveille on the morning of April 10th, the RPO ordered the Company to "fall in and report for muster". Recruits stood to the side of their racks, and called out their rack numbers and names. When the muster got to Brad's rack, there was no response – Brad was officially AWOL (absent without leave) as of that silent moment. In time of war, being AWOL is a felony[12]. The entire Company, including the RPO, was ASTONISHED. Brad had found a way to escape from Shit Island – across the deep water surrounding the compound, to breathe freedom once again, on his own terms.

That same day the Company CPO asked whether any of us knew of Brad's whereabouts. No one had anything to say. Several days later, an NCIS agent stopped by during one of our short breaks after lunch, and talked to recruits that either knew Brad, or were bunk neighbors. Still no

---

[12] See: http://www.public.navy.mil/bupers-npc/reference/milpersman/1000/1600Performance/Documents/1600-010.pdf. Describes AWOL legal implications. Retrieved on Oct. 12, 2016.

story. Then, about a week later, the NCIS agent returned to talk to Dan, only this time the NCIS agent had a letter from Brad, addressed to Dan Sorensen, Charlie Company at RTC San Diego.

"I'm Agent Belisle, Dan, and the purpose of this meeting is to determine your friend Brad Stockley's location. As you know, he's AWOL. He's in deep trouble, so whatever information you can provide might help him out of this jam. Here's a letter addressed to you. We intercepted it with a Court order. Read it and tell me what you think."

"Whoa – Dan says he swam the waterway out of "Shit Island", eluding the sentry, got to the Interstate, and hitched a ride with some hippies in a VW van. They dropped him off in Santa Cruz, and he made his way home from there."

"We know, Dan. The entire letter has been transcribed and entered into Recruit Stockley's case file. We talked to his parents, and they haven't seen him. Chances are he's holed up with a friend. Did he ever talk about a girl friend?"

"No. Brad never said anything about going steady."

"Ok. Any idea why he signs off with "You and Mike can beat the odds, but I just don't belong in that briar patch."? "Who's Mike?"

The NCIS Agent asked me the same questions, and I had nothing. I pointed out that Brad was a decent guy, but he had a bone to pick with the Navy over a breach of contract.

"Breach of contract? He's going to find out real soon that Uncle Sam's enlistment contracts don't have any wiggle room for haggling. We're still doing mining operations off the coast of Viet Nam[13], and no judge is going to look lightly at desertion when the Country's at war."

---

[13] For more info see http://www.vietnamgear.com/war1973.aspx Retrieved Oct. 20, 2016.

I came away with the distinct impression that Brad might have found a way into Canada, like some of his college friends. Well, I thought, it's his life to live. With any luck he won't end up in Leavenworth.[14]

After I graduated from the RTC as a Seaman Apprentice in late May, heading to my "A" School, I spent a few days with Dan Sorensen at his parents place. There was another letter from Brad waiting for Dan's arrival, unopened this time.

"Mike – Brad says he's with his dad. His dad is going to get him off the hook for the AWOL/Desertion charge. The hippies dropped him off in Santa Cruz where he hooked up with a commune for a while. He reached his dad by pay phone and talked him into picking him up to return home."

Rumor has it that Brad beat the rap. Dad piled on some heavy "baksheesh".

I Can't Tell You What the Job is About

Three weeks before graduation a Navy civilian addressed the Company:

"The Navy needs people with foreign language skills, so if you know a foreign language, raise your hand."

About four of us raised a hand.

"Step over here. Fill out these forms."

The forms were to disclose what foreign languages you spoke, your degree of fluency, and how acquired. As soon as we completed them, we were handed a consent form to get tested in Language Aptitude the following day.

---

[14] For more info see https://www.bop.gov/locations/institutions/lvn/ Retrieved Oct. 20, 2016.

I showed up for the Defense Language Aptitude Battery (DLAB) test, which turned out to be based on Esperanto.[15] Sometimes it seemed like the questions were in a language that approximated Latin, and other times the sounds and words really threw you off, like gibberish. Even so, there was a discernible syntax and a grammar structure, so a true linguist would not be stumped. Luckily, with my knowledge of Spanish and French and Latin I was able to cut through the fog, and score well enough to get to an interview with the civilian before the end of the week.

"Have a seat, Makfinsky. You did pretty good on the DLAB. What languages do you have?"

I noticed some diplomas on the wall, one from the Defense Language Institute for Vietnamese. "I'm fluent in Spanish," I answered, "studied French for about eleven years, and took Latin in Elementary School."

"Ok, good enough. It's likely that you would succeed in learning another language, one that the Navy needs for operations. Right at the top of the list is Russian, followed by Korean, Chinese, and Arabic. Do you have skills in any of these?"

"No I don't. I'm not even familiar with their alphabets. My background is in Romance languages."

"Not a problem. I learned Vietnamese, and the only language background I had going in was four years of high school Spanish... more like LA Chicano than Spanish. Operated out of Da Nang, attached to the MACSOG[16]..." He drifted, his thoughts fleeting back to Da Nang... just as quickly, he came back. "But that's max nix. What you need to do is come

---

[15] Posted by: Michael D. Jenning on August 14, 2013 http://dlabprep.com/how-is-the-dlab-test-organized/ Retrieved Oct. 12, 2016.

[16] Plaster, John L. (1974). Secret Commandos: Behind Enemy Lines with the Elite Warriors of SOG. New York: Simon and Schuster. ISBN 0684856735.

up with your language of choice. Keep in mind that depending on your language, your assignment will follow."

Visions of cloak and dagger ops percolated as I considered what each of these languages meant in the context of our foreign policy. "Can I think about it?" I asked.

"Sure – you have all of 24 hours to think about it. Here's today's New York Times; read through it, take a stand on where you fit to move our country forward, and let me know tomorrow. I'll tell your Company's Chief that you need to get back to me by 10 am. Dismissed!"

I left the room, closing the door behind me, and saw another recruit sitting in the hallway. His legs and arms were nervously rigid, bent at right angles.

"No worries, dude" – I said to him – "you're going to hear about some absofuckingly life-changing stuff from that ass-kicker!"

"Cool!" His edginess dissipated.

The door opened. A voice called out - "Recruit Huang, Front and CENTER!"

As I made my way back to the Company barracks, I glanced at the May 3rd, 1974 newspaper the civilian had given me, and an article caught my eye:

Alger Hiss Sees 4 Words in Nixon Transcript as Chance for Exoneration

In an interview with a NYT reporter, Hiss, the man convicted of espionage and treason against the United States of America, who ended up sentenced to five years in prison on two counts of perjury, said that President Nixon, who was obsessed with the Hiss[17] trial, revealed (in one

[17] Joan Brady: America's Dreyfus, Arcade Publishing, New York: 2016. Describes Nixon's case against Hiss.

of his secret Oval Office recordings of a conversation with John Dean) details of FBI operations that proved Hiss was framed by the FBI. It was yet another grain of sand onto the mountain of legal debris from Hiss' ongoing battle for exoneration. Somehow, the thought that this article triggered in my mind was that the Cold War was in fact the most important pillar in U.S. Foreign Policy. My choice of language could be none other than Russian – the official language of the U.S.S.R., dominant as well throughout the entire Eastern Block / Warsaw Pact nations.

The following day I resumed the conversation with the civilian.

"Sir, I've decided on Russian."

"Good decision. It's a difficult language, but you'll have many duty stations to choose from. In fact, you can make a career out of it, and I know many Chiefs (high-ranking Navy non-commissioned officers), and even Master Chiefs (the highest rank among Navy non-commissioned officers, or E-9's) who have specialized in Russian over their twenty-plus years."

"What are some of the duty stations?"

"I can't go into detail, but you will initially be stationed overseas, either in Europe, or in Japan, and chances are that you will serve on any or all types of our Navy platforms, airborne, surface, and subsurface."

'That's good, because I enlisted to become a Sonar Technician, and I was looking forward to Sub duty."

"No doubt you'll get to know some Sonar Techs during deployment, but you'll be assigned to a different rating, a rating called "Cryptologic Technician", or CT for short. I'm enrolling you in the next available Basic Russian Course. It lasts 48 weeks, at the Defense Language Institute in Monterey, California, not too far from here. Once you graduate from DLI, your "A" School phase-2 will be at Goodfellow Air

Base, in San Angelo, Texas. By the time you get out of "A" School you will be a Cryptologic Technician - Interpretive, with the rank of Seaman, or E-3. Stay on the straight and narrow, and you might make Master Chief in record time."

I signed some more papers, and headed back to the Company barracks to catch up with my group. Monterey Peninsula, I thought to myself, home to John Steinbeck's Cannery Row, and I would be there for an entire year! So far, so good – this temporary boot camp blues hiccup is leading to another language under my belt, a "vacation" in Monterey, spook school in San Angelo, and an assignment overseas for air-surface-subsurface cloak and dagger operations. IN LIKE FLINT! - said I. From that point on, boot camp was merely a blur, a speed bump on the road to Cold War operations.

Presidio

She seemed like a Goddess poised at the very top of Presidio – Spanish for vantage point, and I stared at her without waving as I walked up the hill. She wore a deep unabashed Mediterranean tan, showed shapelier than a Greek statuette, and her long, chestnut colored hair blew gently in an ocean-breeze that ascended towards the heights of this vantage point. Her mahogany irises were encircled by snow-white orbs of scintillating energy. To me that energy spelled pure 1974 LOVE.

Victoria had bounded across Ocean and Continent from London, England, where she was undergoing total immersion in English language and culture, to take on an uncertain future with a musician-turned-GI-turning-CTI. BOZHE MOY (*Russian for Good GOD!*) I must have done something right back in Spain – or was it that she had not divined what kind of Cold War adventure our marriage would cast upon her. We got married on September 9th, 1974 in Pacific Grove, California, (40 years later we returned there to celebrate our anniversary on Lovers' Point), and now, 42 years later, neither fortune, nor tragedy, nor victory have blurred the loving spirit of that union. Often the two of us were to reflect on the meaning of the slogan "La Buena Vida NAVY". But I digress...

For most of us it's not easy to learn a new language, especially one with a challenging foreign alphabet such as Cyrillic[18]. The faculty has to be gifted and patient and the students must not only be studious, but seriously goal-oriented in order to overcome the alarming attrition rate. Russian is a "Category III" level-of-difficulty language, and at the DLI that calls for 47 weeks of initial instruction. The more difficult "Category IV" languages include Chinese Mandarin, Japanese, Korean, and Arabic, all with 63 weeks of initial training. When one considers that we English speakers are usually not very effective writers and readers until we reach sixth or seventh grade, a 47-week course (which, by the way, is not a total immersion course of instruction, since classes only go from 08:00 to 16:00 Monday through Friday, with a 1-hour lunch) seems woefully inadequate.

So, what's the DLI formula for success? Opinions vary, but my experience tells me that the faculty skills and instructional materials contribute about 50%, while the Department of Defense (DoD) system of rewards and setbacks yields another 30%, and personal motivation, discipline, and family-friend-peer support fills-in the remaining 20%. For those who are not familiar with the types of rewards and setbacks that the DoD can apply to its enlisted personnel, rewards include favorable performance evaluations that lead to timely promotions and pay increases, whereas setbacks include loss of week-end liberty (up to confinement to the barracks), additional duty after normal work hours,

---

[18] http://learnrussian.rt.com/alphabet/the-history-of-the-cyrillic-alphabet/ Retrieved Oct. 12, 2016. The Cyrillic alphabet owes its name to the 9th century Byzantine missionary St. Cyril, who, along with his brother, Methodius, created the first Slavic alphabet—the Glagolitic—in order to translate Greek religious text to Slavic. It is on the basis of this alphabet that the Cyrillic alphabet was developed in the First Bulgarian Empire during the 10th century AD by the followers of the brothers, who were beatified as saints. Based on the Greek ceremonial script, the original Cyrillic alphabet included the 24 letters of the Greek alphabet and 19 letters for sounds specific to the Slavic language. The Cyrillic alphabet has gone through many reforms in both Russia and other countries. In Russia, the first reformer of the Cyrillic was printer and publisher Ivan Fyodorov.

removal from the class convening and subsequent set-back to a later class convening, which equates to a later graduation date and loss of promotion opportunity. In extreme cases the DoD chain of command can remove a non-performing student from a "Category III or IV" language and enroll him or her in a less difficult course of instruction (e.g. Spanish), or send the student back to his or her parent Service detailer to either "sail", or "fly", or "shoot", all shorthand for being remanded to basic grunt work without any formal technical training for the duration of the drop-out's enlistment contract.

As I said, the faculty and instructional materials constituted the main battery. Faculty members brought skills, worldly experiences, and teaching methodologies that ranged from boldly challenging to utterly confrontational.

"Gospodin *(pron. Gahs-spah-deen, meaning "Mr.")* Barlow – please translate the word "What" into Russian" – commanded the distinguished, yes, professorial Russian instructor.

"Chto", replied Mr. Barlow, only he pronounced it as *"Shhtow"*, letting the final "W" take over, such that the word hung in the air, like a sore *TOE.*

"Nyet, Gospodin Barlow. You fail to notice how concise this spelling is: three simple letters, mind you... Ch – T – O spells SHTO!" he growled with metronomic staccato. **"**Repeat after me: **ShTO!"**

As much as "Gospodin" Barlow tried to emulate Professor Von Damm's diction, his Texas drawl consistently got in the way... the W was there to stay.

As Barlow was being reduced to a squawk-box, one of the senior Petty Officers (Vietnamese linguists being re-purposed into Russian) intervened:

"Professor Von Damm – it will be obvious to any Russian that listens to him that he's Texan, which is kind of like USA's Siberia".

From that point on, the entire class adopted the "**ShTO!**" staccato as an appropriate sound-track to accompany the then fascinating and inimitable Bruce Lee karate chops. We students would break for recess on the sidewalks outside the classroom complexes, toss Frisbees, while simultaneously simulating Bruce Lee karate chops to the sounds of "AIIEEEhh" (preparation of upright chop arm & hand).... "**ShTO!**" as the crushing blow came down upon some simulated hapless student of the Russian language.

And what's so hard about learning Russian? The answer is MANY THINGS: in terms of syntax, one has to contend with the complexities relating to verbs of motion, verbal aspects[19], the declension cases for nouns, etc. In the grammar department, it's imperative to command the rules relating to numbers/dates/numerals. Spelling? One must master the impenetrable art of sprinkling "soft signs" and "hard signs" throughout all of the pronounceable letters. And finally, to make the story short, the *piece de résistance* is its pronunciation anarchy, since there is no straightforward default rule on how to stress polysyllabic words. Spanish, for example, defaults the stress onto the penultimate syllable, and exceptions are clearly marked with "accents" (sure, various

[19] From Linguischtik Wordpress https://linguischtick.wordpress.com/2015/05/03/comparing-the-complexity-of-languages/ Retrieved Oct. 12, 2016. In Russian, aspect distinctions are indicated with verb prefixes. Russian is commonly described as having two aspects, the perfective (complete action) and imperfective (incomplete action). For example the verb читать (*chee-tat*) means 'to read' and has an imperfective interpretation. The verb про-читать (*pro-chee-tat*) also means 'to read', but the prefix про- (*pro*) suggests that the reading is completed. There are numerous prefixes that can indicate perfective aspect, and you need to memorize which verb takes which prefix. Some verbs can take more than one prefix, each of which changes the aspect and adds some new meaning to the verb. The verb писать means 'to write'. It can become perfective with the prefix на-, but also with the prefix за-, in which case it takes on a meaning closer to that of the verb 'to record'. But what if want to have the verb 'to record', but in the imperfective? In that case, Russian grammar has an "infix" -ыв- that you can add back into the middle of the verb root, to make it imperfective again.

different types of accent marks to contend with, but the guidance is plain to see). Russian, on the other hand, leaves you in the wilderness. A babe in the woods, you will surely mispronounce a key word when interpreting for your most discerning and demanding audience. Heteronyms (i.e. same word, but different stress) abound in Russian, so that *zam**O**k* (lock) vs. *z**a**mok* (castle) can trip you up.

You might also choose the wrong meaning for a true homophone (word with exactly the same spelling and pronunciation, having two or more distinct meanings). Take for example the Russian word "brak", meaning variously *marriage* or *defective (item).* Russians will tickle your funny bone with: "Chto-to khoroshyeye ne nazivayut BRAKom", which can be translated as

a. They don't call something good **defective**

b. They don't call something good a **marriage**

Other pitfalls await in terms of false cognates. English speakers hear Russian words that resemble terms that are near and dear to them in English, paving the way to comprehension disconnects. For example, Russian "*ka-bi-nyet*" (phonetic for "office") has us picture a cabinet, or closet, depending on the context, so that when Nikolay says to Jane "Poidyom v moy kabinet" which translates to *Let's go to my office,* Jane mistakenly thinks that she is about to be seduced into a closet *(Let's You and I step into my **kabinet**),* triggering a definite shock to the system in the context of arms control implementation whereby Jane is the arms control inspector at the Votkinsk ICBM production facility[20], and Nikolay the escort at this missile production facility, one which is (was) being

---

[20] U.S. Treaty-Monitoring Presence at Russian Missile Plant Winding Down (Article in NTI Journal, November 20, 2009) http://www.nti.org/gsn/article/us-treaty-monitoring-presence-at-russian-missile-plant-winding-down/ Retrieved Oct 12, 2016 // Votkinsk Machine Building Plant is a machine and ballistic missile production enterprise based in Votkinsk, Russia.

closely monitored by the U.S. Government under the auspices of the Intermediate-range Nuclear Forces (INF) Treaty.

In spite of the complexities, about four fifths of us made it through the DLI 47-Week Basic Russian Course, graduating with flying colors. At the top of the class was a CIA officer who was already fluent in Chinese Mandarin and Cantonese. Perhaps it was he, who single-handedly first and foremost served us greenhorns as a constant source of inspiration and motivation, much more so than the DoD's system of rewards and setbacks.

But it's one thing to have a BASIC understanding of the language, and quite another to be able to apply that basic knowledge to breaking military communications. None of us had any illusions that we would be serving as translators and interpreters for groups of military cadres on "tour" in either country. The prospect of us having to deal face-to-face with Soviets wasn't even remotely realistic, unless we were thrown into some kind of formal interrogation scenario. That kind of linguistic challenge calls for much greater skills than those we had acquired at the DLI basic course. However, because of the clearance levels that we military cryptologic language analysts hold, often times we were called upon to perform face-to-face interpretive work of the cloak and dagger variety.

I point this out because when the now infamous Mig-25 Foxbat fighter aircraft pilot Victor Ivanovich Belenko[21] (yes, of Ukrainian descent) defected to the USA by flying his Foxbat to a base at Hakodate, Japan, on September 6th, 1976, one of my Navy CTI-Russian colleagues served as a translator-interpreter throughout the methodological processes of questioning Belenko about the Foxbat capabilities, and in gaining an understanding of the control switches in the aircraft's cockpit. For that exacting work he was awarded the Navy Commendation Medal. I myself was put to the test in 1998, while serving as the Chief Russian

---

[21] MiG Pilot: the Final Escape of Lt. Belenko, by John Barron, 1980, ISBN 0-380-53868-7.

interpreter/translator for the U.S. Task Group Commander from DESRON-14 (**Des**troyer **S**quad**ron**) during the U.S. Navy and Marine Corps' participation in Operation Sea Breeze 1998.[22] The skills called upon to perform live simultaneous interpretation in formal politico-military settings are, in my opinion, far more demanding than those required to extract valuable information from recorded or even real-time communications. But, once again, I digress.

My much anticipated graduation ceremony at DLI came and went, with VIP's and local politicians congratulating we newly minted linguists. In an unexpected flash of solidarity, my high school friend and old neighbor Dan Waugh decided to pay Victoria and me a visit the week-end after I graduated from Basic Russian. Dan, who, like me also grew up in Madrid, Spain, was one of the key trailblazers who had explained to me the advantages of enlisting in the Navy over blind submission to the

---

[22] http://www.globalsecurity.org/military/agency/navy/desron14.htm. Retrieved Oct 12, 2016. From October of 1998 to July 1999, COMDESRON 14 validated the TACDESRON concept with several exercises, literally around the world. The first was a short notice "fly-away" deployment to the Black Sea where they provided the leadership, warfighting skills and quick response to conduct Exercise Sea Breeze 98. The Ukrainian-US naval exercise "Sea Breeze'98" was held in Ukraine on October 25 - November 4, 1998, on the subject "Carrying out of peacekeeping operations, providing of humanitarian relief and carrying out of search/rescue operations on the territory of a country having suffered from an earthquake." The Sea Breeze'98 maneuvers were held according to the Ukrainian-US cooperation program and in the spirit of the Partnership for Peace Program. NATO members and partners were invited to participate in Operation Sea Breeze '98, with Bulgaria, Great Britain, Georgia, Greece, Italy, Russia, Romania, Ukraine, Turkey and France accepting. Over 30 ships and vessels were involved in the exercise, as well as one submarine, 10-12 planes, and 10-12 helicopters. The exercise's naval phase was held in the Black Sea's northern area, the coastal one was staged on the Shyroky Lan playground near Mykolaiv. During the exercise, an international peacekeeping and humanitarian relief team was formed with a purpose to create a safety zone, deliver humanitarian relief to the population, evacuate patients and wounded. Peacekeeping forces had a task to ensure an embargo on illegal arms supplies to a country dubbed "The Coastal Republic", and to be ready to resort to arms to protect peacekeeping forces.

whims of the Draft Board. Out of active duty, and now a Navy Reservist, he rode his Norton "Commando-750" *classic* motorcycle down from San Jose to our humble 1-bedroom newly-weds' apartment in Pacific Grove. After a leisurely Saturday lunch we took the party to Carmel, and made a stop at Clint Eastwood's Hog's Breath Inn. I wasn't in the least bit surprised about how well Dan, a seasoned sailor, could hold his liquor. Victoria was overjoyed at the opportunity to reminisce with a fellow "Madrileño", and we all conversed till the Hog ran out of breath (an allusion to me, for sure). We ran up a legendary tab - one that Dan, luckily, was able to cover, since dish-washing was out of the question, and my E-2 Seaman Apprentice paycheck had absolutely no silver lining. That soiree was to be our last outing in Monterey during that century, one to remember fondly.

Hog's Breath Inn, Carmel CA, circa 1975

To bridge the gap between knowledge of a foreign language and effective translation of "target" military communications, the DoD has what is known as its "School for Applied Cryptologic Sciences" (SACS) at Goodfellow Air Force Base in San Angelo, Texas.[23] The base hosts a Navy

[23] Manning, Thomas A. (2005), History of Air Education and Training Command, 1942–2002. Office of History and Research, Headquarters, AETC, Randolph AFB,

Detachment that now belongs to the Center for Information Dominance (CID) Command in Pensacola, FL. It's here where Cryptologic Technicians – Interpretive (CTIs) learn Signals Intelligence (SIGINT), and more specifically Communications Intelligence (COMINT) tradecraft. Here is where I learned why we CT's collectively "put miles on the dials", and although today these analog "dial" technologies are a thing of the past (but I'm sooo glad that my car audio system still has control dials!), the digital computer is today what figuratively spins the search dial, stopping on signals of interest that still have to be listened to, graphically analyzed, and assessed for content of interest by crypto-analysts.

The basic elements of COMINT production remain the same, whether the communications targeted are in the Radio Frequency medium, or in the Cyber Space medium. It's essential to determine:

Who the communicating parties are.

Where those parties are located.

The organizational function of the transmitter (sender) and receiver.

Time and duration of communication (*in cyberspace the times of origination/receipt are in microseconds, and duration is not of consequence*).

Frequencies and/or digital networks used and other technical details of note in the transmission.

Encryption or RF scrambling techniques used and whether these can be decrypted or reversed.

Texas ASIN: B000NYX3PC. Under the USAF Security Service, Goodfellow's mission in 1958 became the training of Air Force personnel in the advanced cryptologic skills that the Security Service required. Eight years later, in 1966, the mission expanded further to include joint-service training in these same skills for U.S. Army, U.S. Navy, and U.S. Marine Corps personnel.

Most of the instructors at Goodfellow (affectionately called *Good Buddy* by those of us who have lived the experience) were thoroughly versed in Cold War SIGINT tradecraft. As a matter of fact, my class' Chief instructor, a Senior Chief CTICS who sported a very spiffy set of golden "dolphins" (the Navy's Submarine Warfare insignia), was one fine day pulled from class as we students looked on in awe, and sent out that same night to conduct an extremely sensitive mission series in the Arctic region. Much later I would find out that he was deployed in support of an attempt to enhance Operation Ivy Bells.[24]

Students at the Goodfellow SACS were not thrilled by being boxed-into classrooms with no windows, or by the fact that homework assignments had to be done within the sterile work environment that is a fenced, gated, cipher-lock-restricted "Sensitive Compartmented Information Facility" (SCIF). That said, I admit that from a career standpoint, this SCIF incarceration was adequately compensated by my step promotion from E2 to E3, which was relatively pro-forma with time in grade, completion of knowledge, skills and abilities (KSA) courses, and a clean disciplinary slate.

---

[24] See website http://www.military.com/Content/MoreContent1/?file=cw_f_ivybells. Retrieved Nov. 2, 2016. In October 1971 the United States sent the purpose-modified submarine USS *Halibut* (SSGN-587) deep into the Sea of Okhotsk. Divers working from the *Halibut* found the cable in 400 ft. (120 m) of water and installed a 20 ft. (6.1 m) long device, which wrapped around the cable without piercing its casing and recorded all communications made over it. The large recording device was designed to detach if the cable was raised for repair. The tapping of the Soviet naval cable was so secret that most sailors involved did not have the security clearance needed to know about it. Each month, divers retrieved the recordings and installed a new set of tapes. The recordings were then delivered to the NSA for processing and dissemination to other U.S. intelligence agencies. The first tapes recorded revealed that the Soviets were so sure of the cable's security that the majority of the conversations made over it were unencrypted. The eavesdropping on the traffic between senior Soviet officers provided invaluable information on naval operations at Petropavlovsk, the Pacific Fleet's primary nuclear submarine base, home to Yankee and Delta class nuclear-powered ballistic missile submarines.

Social life during my time at Goodfellow offered much in the way of new experiences. From a climb to my highest point yet by becoming a Dad through my first son Ivan's birth at San Angelo's Shannon Memorial Hospital, where I proudly stuffed Dominican cigars (would have been Cuban, but they were outlawed by State Department embargo) into the hands and pockets of the doctor and his admirable staff, all responsible for the Ivan's smooth delivery, to a series of week-end desert outings with fellow students to hunt for rattle snakes using .22's and cowboy boots (ah, yes, what better a toy for a baby Texan than a real rattler!) San Angelo was all about NEW. Victoria would deliver our other two sons Paul and Daniel during our first Navy tour of duty in Rota, Spain, where we would arrive on Columbus Day in the fall of 1975, after I graduated from the Goodfellow course of instruction, and the three of us took the "return trip" hop across the Atlantic.

How did I luck-out and get orders to Rota, Spain? – in August the CTI detailer at the Navy Bureau of Personnel had let me know, through a Navy Detachment Goodfellow Yeoman, that there were two high-priority overseas assignments for Russian linguists requiring immediate fill: Atsugi Japan, and Rota Spain. My choice was a no-brainer: *stay in warm climes, young man*, my inner voice whispered (while my better half commanded).

I was on a flight to Rota, Spain, only now as an Active Duty Third Class Petty Officer, CTI Russian Linguist in the United States Navy. The NEWNESS was accelerating, and only getting bigger and better. My juvenile enthusiasm was paying off in spades!

II. Where You Go

What defines empire? I used to argue this point with my Spanish friends, who had studied Spain's history, a tremendously rich national heritage which showcases how a country develops imperial aspirations and launches into global hegemony. In the case of Spain these imperial aspirations originated with the 15th Century Catholic Kings Ferdinand and Isabella, who in the course of their lifetimes retook the Iberian Peninsula, expanded Spain's presence in the Western Mediterranean, and began the colonization of America. The history of Spain also dramatically embodies the rise and fall of empire. From Spain's 16th Century imperial apogee in Europe, under Phillip II[25], to its exploration and colonization of America, the Philippines, and places in between, to its final loss of Cuba, the Philippines, and her American and African possessions in the late 19th and 20th centuries. There was a time (almost two centuries) when the Sun never set on Spain.

My friends contended that similarly, the United States of America was now an empire, if for no other reason than because it possesses a multitude of military bases spread across the entire globe. What once applied to Spain can apply today to United States of America. That global military presence equated to, in their minds, the trappings of a dominant

---

[25] Elliott, J.H. (2002). Imperial Spain 1469–1716. London: Penguin Books. ISBN 0-14-100703-6. Under Philip II, Spain reached the peak of its power. However, in spite of the great and increasing quantities of gold and silver flowing into his coffers from the American mines, the riches of the Portuguese spice trade, and the enthusiastic support of the Habsburg dominions for the Counter-Reformation, he would never succeed in suppressing Protestantism or defeating the Dutch rebellion. Early in his reign, the Dutch might have laid down their weapons if he had desisted in trying to suppress Protestantism, but his devotion to Catholicism would not permit him to do so. He was a devout Catholic and exhibited the typical 16th century disdain for religious heterodoxy; he said, "Before suffering the slightest damage to religion in the service of God, I would lose all of my estates and a hundred lives, if I had them, because I do not wish nor do I desire to be the ruler of heretics."

imperial force, one that could control the destinies of smaller nation states.

I argued that the United States was not an imperial nation, and that the military bases were the natural result of defensive wars in which the USA, or our allies, had repelled foreign aggression. My point was that these U.S. military bases (the ones in Germany, Italy, and Japan being punitive "insurance policies" against the WW-II Axis powers, others the result of mutually beneficial bilateral agreements – as in the case of Spain) were established to ensure freedom from aggression, and they are now the visible underpinnings of a global security architecture that benefits not only the USA and Western Europe, but mankind in general. The U.S. military is pre-positioned for preventing the rise of another tyrannical fascist power of the likes of Hitler and Nazism.

But when it comes to the U.S. base in Guantanamo[26], its origins made my point tougher to argue, given that Guantanamo is war

---

[26] Strauss, Michael (2009). The Leasing of Guantanamo Bay. Praeger Security Intl. ISBN 978-0-313-37782-2.

Elsea, Jennifer K.; Else, Daniel H. - Naval Station Guantanamo Bay: History and Legal Issues Regarding Its Lease Agreements (PDF). Washington, DC: Congressional Research Service. Retrieved 9 December 2016. During the Spanish–American War, the U.S. fleet attacking Santiago secured Guantánamo's harbor for protection during the hurricane season of 1898. The Marines landed at Guantanamo Bay with naval support, and American and Cuban forces routed the defending Spanish troops. The war ended with the Treaty of Paris of 1898, in which Spain formally relinquished control of Cuba. Although the war was over, the US maintained a strong military presence on the island. In 1901 the US government passed the Platt Amendment as part of an Army Appropriations Bill. Section VII of this amendment reads: That to enable the United States to maintain the independence of Cuba, and to protect the people thereof, as well as for its own defense, the government of Cuba will sell or lease to the United States lands necessary for coaling or naval stations at certain specified points to be agreed upon with the President of the United States. After initial resistance by the Cuban Constitutional Convention, the Platt Amendment was incorporated into the Constitution of the Republic of Cuba in 1901, which went into effect in 1902. Guantanamo Base stood up in 1903.

reparation for a conflict that apparently the USA instigated to trigger the infamous 1898 Spanish-American war.[27] The extremely ambiguous circumstances surrounding the sinking of the USS Maine in Havana Harbor on February 15th, 1898, which motivated the U.S. declaration of war on Spain, can be seen as pretext for imperial expansion.

Agreement providing conditions for the lease of coaling or naval stations, signed 7-2 1903

The 1903 lease agreement was executed in two parts. The first, signed in February, included the following provisions:

1. a promise by Cuba to lease to the United States a specified area at Guantanamo Bay "for the time required for the purposes of coaling and naval stations";
2. the right to acquire any privately owned land within the leased area "by purchase or by exercise of eminent domain with full compensation to the owners thereof";
3. the right to use the areas as naval stations, and for no other purpose, with a non-exclusive easement to adjacent waters;
4. consent on the part of Cuba to the US exercising "complete jurisdiction and control over" within the leased area;
5. recognition by the US of Cuba's "ultimate sovereignty" over the leased area.

The second agreement, signed five months later in July 1903, established the amount of USD$2,000 to be paid to Cuba annually by the US. Additional stipulations included the following:

1. payments were to be made in gold coin;
2. the US would pay to build and maintain fences marking the boundary of the leased area;
3. commercial and industrial activities in the area would be restricted;
4. mutual right of extradition
5. a duty-free zone, but not a port of entry for weapons or other goods into Cuba proper
6. Cuban shipping to have the right of access to the Bay
7. ratification to be within seven months.

The payment of $2,000 was increased to $4,085 in 1934. Only one lease payment has been cashed since the Cuban Revolution.

[27] Tucker, Spencer, ed. (2009). "Chronology". The Encyclopedia of the Spanish-American and Philippine-American Wars: A Political, Social, and Military History. Santa Barbara, Calif.: ABC-CLIO. ISBN 9781851099511.

I arrived in Spain with Victoria and Ivan in the fall of 1975, while Generalissimo Franco was on his deathbed. He had fallen into a coma on October 30th, and passed away on November 20th, after his family consented to removing all life support systems. Francisco Franco de Bahamonde was immediately succeeded by King Juan Carlos, Franco's chosen heir to re-instate the Spanish Monarchy.[28] The political turmoil during this transition period presaged radical societal and political changes in a country that had been torn asunder by civil war, and governed through dictatorship for thirty six years. Franco's death was followed by an entire month of mourning orchestrated through the national media. Spanish radio and television continuously and exclusively aired funeral dirges and catholic masses. Throughout his lying in repose at the Royal Palace in Madrid, and his funeral procession at the Valley of the Fallen, a Spanish Civil War memorial West of Madrid, the media produced solemn broadcasts of religious liturgies and political homilies.

I can assure you that there was absolutely no partying going on in Spain as Victoria, Ivan, and I settled into our new quarters at the "Virgen del Rocio" (Virgin of the Dew, *dew being essential to Andalusian agriculture*) Apartment complex facing the sandy beaches of "Playa de la Costilla" in Rota[29].

---

[28] http://countrystudies.us/spain/25.htm Retrieved November 22, 2016. The democratization that Franco's chosen heir, Juan Carlos, and his collaborators peacefully and legally brought to Spain over a three-year period was unprecedented. Never before had a dictatorial regime been transformed into a pluralistic, parliamentary democracy without civil war, revolutionary overthrow, or defeat by a foreign power. The transition is all the more remarkable because the institutional mechanisms designed to maintain Franco's authoritarian system made it possible to legislate a democratic constitutional monarchy into existence.

[29] http://rota.com.es/turismo/tur0quevisi0playas.htm. Retrieved November 22, 2016 – See some of the best Spanish Atlantic beaches.

At that time, in 1975, the U.S. presence in Spain was the result of a bilateral agreement negotiated with Franco in 1953, whereby the USA leased four major bases: Torrejon (a strategically important air base near Madrid as a forward-base for USAF's 16th Air Force[30] – and where I attended High School - GO Knights!), Rota (the deep water naval base near Cadiz, capable of harboring a Carrier Battle Group and replenishing nuclear submarines in its sheltered bay West of Gibraltar), Zaragoza (an air base northeast of Madrid), and Moron (an air base near Seville, capable of bedding down B-52s bombers and KC-135 tanker/refueling aircraft from the Strategic Air Command)[31]. Perhaps to gain ownership of the Spain-USA bilateral relationship, King Juan Carlos ratified a new Treaty of Friendship and Cooperation in January 1976.[32] This new treaty

---

[30] https://hoseyfiles.files.wordpress.com/2015/03/20040501-16af-heritagepam.pdf. Retrieved Nov. 22, 2016.

[31]http://ips.sagepub.com/content/early/2010/08/10/0192512110372975.full.pdf Retrieved Nov. 22, 2016. Over the course of the United States military presence, the status of the United States bases as a domestic political issue varied during the five distinct phases spanning: (1) the era of Franco; (2) the initial transition (1975–1980); (3) democratic consolidation and Socialist dominance (1980–1989); (4) the post-Cold War period (1989–2003), which saw a second alternation in power; and (5) the PSOE's re-election in 2004. Of these eras, the basing issue was at its peak in Spanish politics during phase 3, during which the issue played an important role in two election campaigns for the PSOE and was strategically entangled by the government into the 1986 NATO referendum.

[32]http://ips.sagepub.com/content/early/2010/08/10/0192512110372975.full.pdf Retrieved Nov. 22, 2016. The transition to democracy in Spain began shortly after Franco's death of natural causes at the end of 1975. For a period of six months, little changed: King Juan Carlos took over as head of state, as envisaged by the constitutional arrangements laid down in the 1969 Succession Law, and the authoritarian government of Carlos Arias Navarro was maintained, with small changes. This government quickly concluded the 1976 basing agreement (Treaty of Friendship and Cooperation, January 1976), which proved uncontentious, maintaining the United States military presence along similar lines to the previous two decades, with a slightly improved economic package, and some concessions on the use of Rota by nuclear submarines. However, the increasing levels of social conflict forced a change of government in July 1976, with the

included provisions for harboring U.S. nuclear submarines at Rota Naval Base, a bold step given the fact that Spain had experienced a jolting nuclear weapons accident by U.S. aircraft over the coast of Almeria just 10 years earlier, in January 1966:

"During a routine refueling operation on 20 January 1966 that a B-52 carrying nuclear weapons collided with a KC-135 tanker in mid-air. Both aircraft crashed near the small village of Palomares on the southeastern coast of Almeria, Spain, and U.S. forces made an immediate effort to carry out recovery and cleansing actions. The ensuing operations, directed by the Sixteenth Air Force commander, to recover radioactive material and the "lost" nuclear weapon from the sea gained worldwide attention and profoundly affected U.S.-Spanish long-term relations". For a dramatic re-enactment of the nuclear bombs salvage operations, I highly recommend you enjoy the movie "Men of Honor". It portrays Navy Master Diver Carl Brashier heroically leading the Palomares clean-up. [33] For those historians reading this, I must point out that so far

---

appointment of Adolfo Suárez, who became the architect of democratic transition.

[33] Pringle, Capt. Shuan (February 21, 2001). "Direction, Discipline, Determination: The Story of Carl Brashear". Air Force Space Command, United States Air Force. Retrieved 2008-08-17. In January 1966, in an accident now known as the Palomares incident, a B28 nuclear bomb was lost off the coast of Palomares, Spain, after two United States Air Force aircraft of the Strategic Air Command (SAC), a B-52G Stratofortress bomber and a KC-135A Stratotanker aerial refueling aircraft, collided during aerial refueling. Brashear was serving aboard USS *Hoist* (ARS-40) when it was dispatched to find and recover the missing bomb for the Air Force. The warhead was found after two and a half months of searching. For his service in helping to retrieve the bomb, Brashear was later awarded the Navy and Marine Corps Medal – the highest Navy award for non-combat heroism. During the bomb recovery operations on March 23, 1966, a line used for towing broke loose, causing a pipe to strike Brashear's left leg below the knee, nearly shearing it off. He was evacuated to Torrejon Air Base in Spain, then to the USAF Hospital at Wiesbaden Air Base, Germany; and finally to the Naval Hospital in Portsmouth, Virginia. Beset with persistent infection and necrosis, his lower left leg was eventually amputated. Brashear remained at the Naval Regional Medical Center in Portsmouth from May 1966 until March 1967

there have been approximately thirty two nuclear weapons accidents worldwide, known within DoD circles as "Broken Arrows".[34]

1776 – 1976

While Spain was on its road to revamping its democratic ways, my country was celebrating its TWO-HUNDREDTH ANNIVERSARY of democratic Independence from what was once the British Empire.

To commemorate this historic occasion, Spain sent its celebrated tall ship, the ***Juan Sebastian de Elcano*** to New York's Bicentennial Parade of Square-rigger-Windjammer Tall Ships, the top-drawer flagship event during Fleet Week. We Rota-based sailors had seen the *JSE* depart from the port of Cadiz, then cut across the bay while smartly deploying full sail

---

recovering and rehabilitating from the amputation. From March 1967 to March 1968, Brashear was assigned to the Harbor Clearance Unit Two, Diving School, preparing for return to full active duty and diving. In April 1968, after a long struggle, Brashear was the first amputee diver to be (re)certified as a U.S. Navy diver. In 1970, he became the first African-American U.S. Navy Master Diver, and served ten more years beyond that, achieving the rating of Master Chief Boatswain's Mate in 1971. Brashear was motivated by his beliefs that **"It's not a sin to get knocked down; it's a sin to stay down" and "I ain't going to let nobody steal my dream"**.

[34] http://www.atomicarchive.com/Almanac/Brokenarrows_static.shtml Retrieved Nov. 22 2016 *(note: this archive only lists USSR and US incidents – there could well be many more attributed to other nations with nuclear weapons capabilities)* In the case of the Palomares incident, the Atomic Archive.com says: Date: January 17, 1966 Location: Palomares, Spain / A B-52 carrying four nuclear weapons collided with a KC-135 during refueling operations and crashed near Palomares, Spain. One weapon was safely recovered on the ground and another from the sea, after extensive search and recovery efforts. The other two weapons hit land, resulting in detonation of their high explosives and the subsequent release of radioactive materials. Over 1,400 tons of soil was sent to an approved storage site.

(20 in all) amidst local fanfare, and head West into the Atlantic on its way to New York for Operation Sail (Op-Sail '76).[35]

Later in life I would hear one of my friends in the Spanish Navy brag about it.

"You had to be there! This was Elcano's 48th training cruise. We sailed from Cadiz to arrive in New York for the 4th of July Bicentennial celebrations as one of 12 ships participating in Operation Sail. We made stops in Tenerife, Bermuda and Newport along the way. Those port calls were all unforgettable, but the stay in New York City was to DIE FOR!" "Hell *(actually he used another Spanish expletive which doesn't translate as well: "JODER")*, the fireworks alone were the absolute best that I'll ever see in my lifetime!"

[35] http://www.navy.mil/ah_online/archpdf/ah197601.pdf. Retrieved Nov. 27, 2016. Contains photographs of all the Tall Ships that participated in Operation Sail 76 (Op-Sail '76).

Also in celebration of what America had to offer in the way of political, media, industry, and culturally emblematic personalities, a famous photog of the time, Richard Avedon, sold a series of 69 portraits of socio-politically relevant Americans to Rolling Stone magazine.[36] I remember picking up a copy of it at the "Stars and Stripes" bookstore on Rota Naval Base, and flipping though it to see portraits of Jimmy Carter, Gerald Ford, Cesar Chavez (Organizer, United Farm Workers), and Leonard Woodcock (President, United Automobile Workers), among others. It was graphics and articles like this that helped set the tone for the upcoming presidential elections in which Jimmy Carter – a little known "reformer" beat our incumbent President Gerald Ford by 57 electoral votes (12 percent advantage), but by only 2 percent of the

---

[36] https://iconicphotos.wordpress.com/2012/05/18/the-family-1976-richard-avedon/ Retrieved Nov. 27, 2016. Avedon traversed the country from migrant grape fields of California to NFL headquarters in Park Avenue and returned with an amazing portfolio of soldiers, spooks, potentates, and ambassadors that was too late for the bicentennial but published in *Rolling Stone*'s Oct. 21, 1976, just in time for the November elections. Sixty-nine black-and-white portraits (seen all together in a Metropolitan exhibit) were in Avedon's signature style — formal, intimate, bold, and minimalistic. Appearing in them are President Ford and his three immediate successors — Carter, Reagan, and Bush. Other familiars of the American polity such as Kennedys and Rockefellers are here, and as are giants who held up the nation's Fourth Pillar during that challenging decade: A. M. Rosenthal of the *New York Times* who decided to publish the Pentagon Papers, and Katharine Graham who led Woodward and Bernstein at *Washington Post*. Their source, Deep Throat, is here too: W. Mark Felt, the former associate director of the FBI, although he didn't reveal that fact until 2005 — the year after Avedon himself died. It is also clear here that apart from a few civil rights leaders and eminent wives, the pantheon of 1976 was mostly white, mostly male, and mostly elderly. Yet, some familiar contemporary names amongst its younger members — the activist Ralph Nader, 42; Jerry Brown, 38, then as now the governor of California; Donald Rumsfeld, 44, then and future Secretary of Defense — also suggest this group's political endurance and Zeligian relevance. Consciously or otherwise, absent were the Supreme Court justices and the man whose resignation made this portfolio possible. Instead, Avedon convinced Nixon's secretary, Rose Mary Woods to pose for him.

popular vote in a 56 percent voter turnout of just under 80 million American voters.

The media had impressed upon We the People how Nixon's internal politics called for a reformed system of government. President Ford's greatest liability was that he not only carried Nixon's political baggage into the election, but he, the incumbent, had not been elected President. He was in fact the first person appointed to the vice presidency under the terms of the 25th Amendment, following the resignation of Vice President Spiro Agnew on October 10, 1973. The following year, upon Nixon's resignation (to preclude an impeachment ordeal, which was trumped by a "Presidential Pardon"), Ford took over the Presidency on August 9, 1974 becoming the first and to date the only person to have served as both Vice President and President of the United States without being elected to either office.

As a newly minted military professional I focused on these two candidates' military backgrounds, however. President Ford had been a highly decorated WW-II Naval Officer from 1942 to 1946, serving aboard the USS Monterey aircraft carrier during numerous Carrier Strikes in the Pacific Campaign, to include actions on Wake Island and Leyte Gulf. Candidate Jimmy Carter, on the other hand, was admitted to the U.S. Naval Academy in Annapolis Maryland in 1943. He graduated and later qualified as a Submarine duty officer aboard the USS Barracuda, eventually, by 1952 working his way into Admiral Rickover's nuclear submarine program. He contributed to the design of the USS Seawolf, one of the first nuclear-powered submarines in the U.S. Navy, which boasted a liquid-sodium cooled epithermal reactor, giving a 40 percent reduction of the size of machinery in the engineering spaces when compared to the USS Nautilus, its predecessor. We sailors knew the prestige that service in the "nuclear Navy" meant, and to us Jimmy Carter was a really, ***really*** smart guy.

The Rota Naval Station celebrated our Bicentennial in style. Security was not as much of a concern back then (sure, the ETA Basque separatist organization was on the prowl, but not in Rota), and when the Naval Station hosted an Open House, it was an all-stops-out command performance, with USO-sponsored bands blaring over loudspeakers set up on a stage under the 500,000 gallon red-and-white-checkered water tank.

Amid rock 'n roll, fireworks, Budweiser and Cruz Campo brews, bathed by the light of a full moon and wrapped up in balmy weather forever, we locals, to include American and Spanish *"Roteños"* partied hardy until the wee hours of the morning of July 5th. All was well at this strategic gateway onto the Mediterranean.

Base Naval de Rota

Perhaps the most emblematic artifact at Spain's Rota Naval Station, one that represents the Base, is the TETRAPOD.[37] Arguably the base would not have been possible without this basic building block. The U.S. Navy's historical archive, maintained by CNIC (Commander, Navy Installations Command) states that "Some 10,000 concrete tetrapods, resembling large jacks, were carefully placed *(in the Bay of Cadiz)* to provide a seawall to protect a large artificial harbor".[38]

---

[37] http://www.stripes.com/news/strategic-naval-installation-under-construction-at-rota-1.48094. Retrieved Nov. 27, 2016.

[38]http://www.cnic.navy.mil/regions/cnreurafswa/installations/ns_rota/about/history.html. Retrieved on Nov. 27, 2016. Rota was established in 1953, following the signing of an agreement for facilities use between the United States and Kingdom of Spain. The agreement required two years of surveys, negotiations and planning which led to ground breaking on the base in 1955. Naval Station Rota is located on the Bay of Cadiz between the towns of Rota and El Puerto de Santa Maria. Four entry gates - Rota, Jerez de la Frontera, El Puerto and Fuentebravia - are operated by Spanish security forces. Security inside the base is provided by both Spanish and U.S. Navy security teams. The Commander, Naval Activities (COMNAVACT) Spain is headquartered in Rota and serves as the area coordinator for all U.S. Naval Activities ashore in Spain and Portugal. COMNAVACT Spain also serves as the commanding officer of Naval

An edition of “Stars and Stripes” that came out during the 1950’s construction of the base says this about the ubiquitous tetrapod:

“Queer contraptions which originated in the brains of Frenchmen are being fabricated and dumped into the ocean at Rota to make the Atlantic more pacific.

Constructed of concrete and resembling jacks which might have been used by a young Paul Bunyan, these odd-appearing objects are called tetrapods. They are built on the site and deposited in the water washing the projected 3,000-foot marginal wharf that forms an artificial-harbor in the outer reaches of the Bay of Cadiz.

Nearly 10,000 of these "schmoos," as Gillette *(Rear Admiral Norm Gillette, at the time Rota NAVSTA Commanding Officer, CAPT Norm Gillette)* calls them, ranging from 8 to 25 tons each will be used to break up wave action and absorb a maximum of the ocean's eroding energy.” As the tetrapods erode, the harbor is dredged, and new ones are placed along the breakers.

Station Rota. The commander reports directly to Commander, Navy Region Europe, Africa and Southwest Asia, located in Naples, Italy.

At the time of construction, CAPT Norm Gillette's Executive Officer was Commander Calvin Durgin Jr., another intrepid Naval Aviator who, like Norm, had cut his aviation teeth on aircraft carriers in WW-II Pacific and Atlantic campaigns. Cal was of the opinion that the best way to ensure mission success at Rota was to build a strong, enduring relationship with the Spanish Navy *(Armada, in Spanish)*, so he became one of the co-founders of a social organization called the "Club Hay Motivo", which translates to *"There's a Motive Club"*.

As the story goes, the clubhouse is situated on the site where the Spanish-American teams signed the agreement to begin construction of the Naval Base in 1955. The Spanish Admiral in charge of the project was known to have said "Siempre hay motivo para brindar" *(there is always a reason to toast)* at the conclusion of the agreement. On that positive note, Cal stood to toast in his best, newly learned Spanish:

"Yo brindo para que tenemos mas *meeting*s como este, y quiero que sea nuestro Club Hay Motivo." – Roughly translated, Cal toasted to having more such meetings under the auspices of Club Hay Motivo. Needless to say, Spanish Sherry easily became the CHM's lifeblood.

The CHM has endured the passing of time, and just recently the Spanish "Diario de Cadiz" newspaper published this about the Club: The Club Hay Motivo continues to strengthen ties between the two militaries, although its best years to date were in the 70's and 80's, when Friday plenary meetings were well attended, often giving way to Saturday socials. The most important of these were presided by Don Juan de Borbon *(King Juan Carlos' father, who in 1977 abdicated his right to the throne and returned to Spain from exile... he was named honorary Admiral of the Spanish Armada on July 8th, 1978)*, who deeply appreciated the Country's naval heritage.

Cold War Operations

From a maritime standpoint, the importance of overseas basing is just as much about having forward locations for operating weapons platforms for surveillance and control global of sea lanes (or lines) of communication (SLOCs), as it is about having access to sources of strategically and tactically important intelligence information.

And what kind of advantage did Rota give at the time? For one, in addition to providing a forward basing capability to launch and monitor U.S. operations in the Mediterranean, it was a useful High Frequency radio wave (HF band, comprising the 3-30 MHz band of the radio frequency spectrum, also known as the "decameter" band) collection point.

In the 70's and 80's, before satellite communications became the preferred means, a large share of military command and control communications were carried out in the HF spectrum. Military communications operators could chose to transmit on the HF carrier frequency itself, or they could split the signal into upper and lower sideband, and they could also opt to scramble their information onto the carrier radio waves using complex modulation techniques. The kinds of information that got modulated onto these HF carriers could be voice (employing Single Side Band – SSB), Morse code (manual or automatic... we referred to these as "Manuel" or his close cousin "Otto"), or other more sophisticated machine-to-machine data transmissions.

Perhaps one of the most useful characteristics of HF transmissions, from a Rota standpoint, is that HF signals "bounce" off the Earth's Ionosphere, where they get reflected back onto the surface. This phenomenon is known as "skip" or "sky wave" propagation. Distant HF signals can bounce several times across the globe. That's why in Rota it was possible to receive Chinese and even Soviet Far East (e.g. from the Kamchatka military complex – an important nuclear submarine staging base, among other things) HF transmissions.

The bounce patterns were predictable, although they could be altered by Sun storm activity and other atmospheric and magnetic variables.[39] Rota, in other words, was a suitable collection point for HF signals from Europe, Asia, and Africa. For that reason, in terms of Navy Cryptologic personnel, Rota was one of the largest concentrations of the most widely-varying cryptologic skill-sets: Interpreters for European, Middle East, and North African languages; Morse Operators; Special Signals Operators (collecting machine-to-machine data transmissions in a myriad of modulation formats), as well as Direction Finding (DF) Operators who employed the vast "Wullenweber" circularly disposed antenna array (CDAA) to determine not only the direction that a signal was coming from, but through triangulation with other similar arrays throughout the world, and overhead satellite systems, the probable geolocation of the transmitter. Rota was in effect an agile forward-based intelligence production node, one with an extremely capable High Frequency Direction Finding (HFDF) system nicknamed "BULLSEYE".

As a new arrival, I had not yet volunteered for any of the more dangerous assignments that required qualifying for airborne duty or submarine duty. I could man signal access workstations (called "posits") that were either land-based or shipboard.

"Third Class Makfinsky – I'm going to assign you to take one of the "TEE-BOW" posits. Do you think you can handle it?"

[39] See https://www.arrl.org/files/file/Technology/tis/info/pdf/8312011.pdf for information on HF radio wave propagation. Retrieved on Nov. 27, 2016.

Gunnery Sergeant "Gunny" Cranston was a thin, slow-talking Buckeye. His boss, a Navy Warrant Officer was fond of needling him:

"Q: How do Buckeyes' brain cells die? A: Alone." The Warrant would often start the watch with that quip, setting the tone for Gunny's interaction with the rest of us watch standers. The motivation behind the jab was not clear to us – perhaps a power play, or maybe insecurity over *"dominium feminae"* in that there were many eye-popping barmaids outside the main gate, and the few that spoke English were in high demand.

"Not only can I handle it, Gunny – I can red-line that TEE-BOW if it's Soviet Navy or Long Range Aviation intel we need to wring out of it." I was vaguely familiar with what kind of signals were exploited on a TB posit, which stands for "transient burst" types of signals, because one of the guys on watch had told me about how important these burst signals were to track the whereabouts of Soviet nuclear submarines and long range bombers.

The Rota TB posits were an offshoot of the 1960's Project BORESIGHT[40], a US-Canada burst signal geolocation system that gave JFK

---

[40] http://jproc.ca/rrp/rrp2/boresight.html. Retrieved on Nov. 27, 2016. "One of the prime reasons Khrushchev was ready to capitulate to Kennedy was based on the successes of the Atlantic HFDF Net, specifically with a project known as BORESIGHT, especially the wideband capabilities of these sites. In the exchanges between Khrushchev and Kennedy, Khrushchev threatened Kennedy with the information that the Soviets had nuclear submarines sitting on the bottom of the Atlantic from Florida to Halifax. These were armed with nuclear missiles targeted against major and critical sites such as Norfolk, Boston, Cape Canaveral, New York and Halifax amongst others.
Because the Soviets had been using super burst transmissions, they were obviously convinced that those transmissions could not be DF'd or fixed. Kennedy had been well briefed and was prepared for just such an eventuality and was willing, under these severe circumstances, to take the chance of blowing the cover of wideband direction finding successes. He advised Khrushchev that he was well aware of the disposition of Soviet submarines and then proceeded to provide him where they were located by latitude and longitude. He also advised him how often they reported and that they were all targeted with US Polaris

the upper hand over Khrushchev during the Cuban Missile Crisis. BORESIGHT/BULLSEYE had pinpointed the locations of Soviet nuclear ballistic-missile submarines throughout the Atlantic, and when Khrushchev heard JFK say that these would be destroyed unless the Soviet Union agreed to dismantle the Cuban threat, a deal was struck.

I was given the TB-004 posit (above 007 status!), and was to side-saddle with a CTT (Cryptologic Technician, Technical – this rating concentrated on special signal collection, primarily "bauded" data transmissions in all of the various modulation and enciphering and encryption formats). CTT 2 Urruizaga was all business when it came to recognizing "scratches" (i.e. the sound of burst transmissions) within the RF spectrum. He had a training tape chock-a-block full of "scratches" that he made me listen to for an entire 8 hour watch. Once I got trained in "scratch" recognition, it was my job to collect reports from the Morse signal collection shop (manned by CTRs – Cryptologic Technician Reporter), and from these collate a list of HF Frequencies that the Soviet Navy had selected that day for burst signals transmission. CTT2 "U" would then plug them into his banks of receivers, and monitor them like a hawk. As soon as a "scratch" tripped one of his receivers it would automatically record on a special wide-band recorder. He would play-back that wideband recording into a sonogram to analyze the baud rate and signal characteristics.

Meanwhile, the burst signal of interest was automatically "collared" by the BULLSEYE HFDF system for geolocation, and a triangulated location was provided by the BULLSEYE operator. Once the signal was classified, geo-located, and sonogram'ed, it was reported into

---

missiles. Kennedy also indicated his military were quite aware of the location of the five diesel electric Foxtrot submarines enroute to Cuba and the Cuban submarine facility under construction at Cienfuegos, Cuba. He further informed Khrushchev, that in an instant of time, the first target of US forces would be the destruction of the Soviet submarines, both nuclear and diesel electric. The complete scenario then changed and Khrushchev backed down and commenced to negotiate a deal on the de-installation of US missiles in Turkey.

the world-wide tracking system for the benefit of decision-makers. Follow-up reports would contain any additional information about the transmitting platform (usually a submarine, or a bomber aircraft, although trains shuttling concealed nuclear ballistic missile launching systems might also come out of the woodwork) that might be gleaned from an accompanying voice transmission. You guessed it – it was my job to transcribe and translate any and all of those associated voice cuts. All too often the voice signal had not recorded with any degree of fidelity (not center-tuned), but when it was clear, the information I extracted was usually "Op-Chatter" (operator chatter) having to do with adjustments resulting from equipment troubles, projected outages, or a forecast of next communications window. None of that was challenging from an interpretive standpoint, and while I knew that this was important work, after a few months of TEE-BOW watch-standing, I was ready to move on.

Soviet Fifth Ehskadra (5th-E) – A Force to Recon With in the "Middle Earth's Sea"

Also known as the Fifth Mediterranean (Russians know this Sea as the "Middle Earth Sea" - *Sredizemnomorye*) Squadron of Warships, in the 1970's and 80's it comprised a flotilla of Soviet vessels that included submarines, guided-missile cruisers, destroyers, helicopter carriers, amphibious assault ships, intelligence collection ships, and refueling-resupply-repair ships, not to mention the many afloat barracks facilities that the 5th-E used throughout its permanent Mediterranean anchorages.[41]

This formidable war machine came about thanks to the USSR's then Chief of Naval Operations, Fleet Admiral Sergei Gorshkov[42], a master

---

[41] The Soviet Presence in the Mediterranean – RAND Corporation. Retrieved on Nov. 27, 2016.
https://www.rand.org/content/dam/rand/pubs/papers/2008/P7388.pdf

[42] http://www.airpower.maxwell.af.mil/airchronicles/aureview/1982/jul-aug/chipman.html Retrieved on Nov. 27, 2016. Excellent source for insight into how Gorshkov's planning gave the Soviet Navy its global reach by the 1980's. For

maritime strategist who took the Soviet Navy from a homeland defense force to a global power-projection naval presence capable of challenging the United States Navy in all oceans, as well as in the air and in space. In 1967 Adm. Gorshkov got the Politburo to stand-up the 5th-E as a fully operational semi-autonomous organization – semi-autonomous because its ships would operationally "in-chop" from either the Black Sea Fleet, the North Fleet, or from the Baltic Fleet. It was divided into six operational elements:

Element 50 – Command Ships and Battle Ships

Element 51 – Submarines (averaging 6-8, occasionally more)

Element 52 – Attack Missile-Artillery Ships

Element 53 – Anti-Submarine Warfare Ships

Element 54 – Amphibious Landing Ships

Element 55 – Supply Ships

In the big picture it was Element 51 that held the trump card as Gorshkov's forward Anti-Carrier-Warfare and Anti-Submarine-Warfare strike force. The USSR did not have the resources, or the know-how to employ Carrier Battle Groups (nor did any other Sea Power, for that matter), so they relied on their submarine force to counter the tremendous strategic weight that the U.S. Navy's Carrier Battle Groups (CVBG) and Nuclear Attack Submarines (SSN) and Nuclear Ballistic Missile Submarines (SSBN) brought to American diplomacy and SLOC control. As time went on, and the Soviet Navy put more capable surface warships to

---

an outstanding compendium of Soviet VMF (military fleet) platforms, capabilities, purposes, and intentions during the Cold War, see "Understanding Soviet Naval Developments" released by the Office of the CNO, Director of Naval Intelligence, 3rd Edition Jan 1978 at link: https://babel.hathitrust.org/cgi/pt?id=mdp.39015074796007;view=1up;seq=7. Retrieved on Nov. 27, 2016.

sea, such as the Kresta-II, Kara, Sovremenny[43], Udaloy, and Slava Guided Missile Cruisers, and more capable Long-Range-Aviation bombers such as the Badger, Blinder, Backfire and Blackjack[44], Gorshkov's dream of "simultaneously synchronized sub-surface-launched – surface-launched – and air-launched" Anti-Carrier and Anti-SSBN/SSN strikes became possible. The intent was to launch waves of guided cruise missiles (be they high-explosive or nuclear armed) to overwhelmingly swarm high-value targets (i.e. aircraft carriers, SSN submarines armed with Tomahawk cruise missiles, and SSBN submarines armed with Trident missiles). This tactic represented the epitome of Gorshkov's famous moniker: "Good enough is the enemy of better".

The key to synchronizing this kind of strike was very precise (and survivable) command and control for directing the widely-dispersed team of submarines, guided-missile cruisers, and bombers into extremely

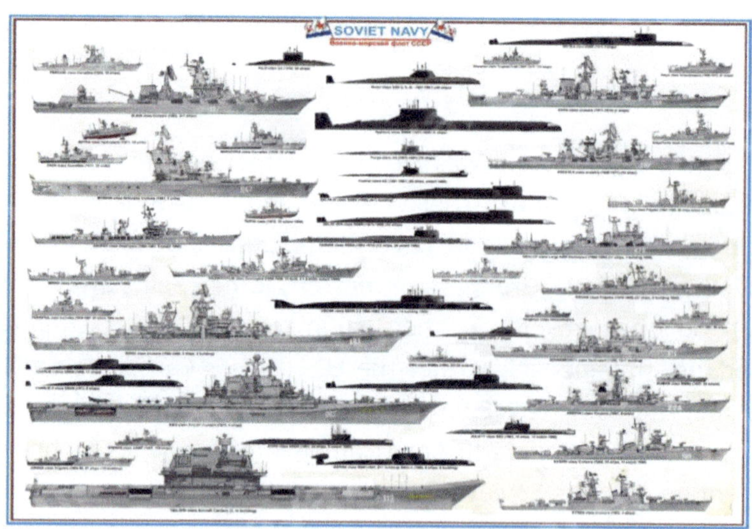

[43] http://www.globalsecurity.org/military/world/russia/956.htm. Retrieved Dec. 2, 2016. The Sovremennyy is capable of striking several surface targets simultaneously.

[44] http://www.military-today.com/aircraft/tupolev_tu160_blackjack.htm Retrieved Dec. 2, 2016. At the time of this writing, the Blackjack is the world's largest operational bomber aircraft, one that has set 44 world records.

complex operations. To prove and ensure that this anti-carrier and anti-SSBN/SSN strategy was feasible, Gorshkov's Navy had to carry out operational strike exercises, and for this the Soviet Navy conducted what our Fleet Ocean Surveillance Information Facility (FOSIF) analysts called "mirror-image Anti-CVBG war games." The submarines and guided-missile cruisers would simulate launching missile salvos against an "exercise" target at sea, while the bombers would conduct synchronized missile launches at an "exercise" target over land. The mirror-image component of this operation revealed itself when FOSIF analysts super-imposed both strike profiles to prove how these two strike exercises were related not only in time, but in that they employed a common simulated target.

In the fog of war, however, many factors can influence the outcome, and it was widely accepted that in order for the Soviet Navy and Long Range Aviation to be successful in destroying a CVBG, an SSBN, or an SSN, they would require either absolute serendipity[45], or, even more unlikely, a completely blind and supremely dumb victim.

Eternal Vigilance is the Price of Liberty – Says the SANDEMAN

Blind and dumb? Gorshkov knew the odds were slim to none. After WW-II, the U.S. Department of Defense built the most capable world-wide surveillance and intelligence production network that civilization has ever seen.

At the Rota Naval Base, the "gateway to the Mediterranean", this U.S. surveillance and intelligence production network was well represented. In terms of surveillance resources we had VQ-2 – Fleet Air

---

[45] Serendipity is postulated by Napier and Vuong (2013) as a 'strategic advantage' with which a firm (*in this case a military organization)* can tap its potential creativity.

Reconnaissance Squadron Two (a.k.a. FAIRRECONRON-TWO[46] with both land-based EP-3E Aries and carrier-based EA-6B Skywarrior Signals-Intelligence /SIGINT/ aircraft), a VP – Maritime Patrol aircraft Detachment (with P-3B, later P-3C Orions to conduct not only Magnetic Anomaly Detection (MAD) and optical-visual reconnaissance, but anti-ship and anti-submarine warfare aggressor missions as well), FOSIF Rota with its Tactical Support Center (TSC), and an Anti-Submarine Warfare Operations Center (ASWOC). In terms of Intelligence production resources, Rota's FOSIF and the Naval Security Group Department (NSGD – with a reach into overhead satellite sensors and "National" threat information streams) were the main battery, while commands such as the Meteorological and Oceanographic (METOC) Command and the Naval Communications Station (NCS) surveilled weather conditions and radio-wave propagation characteristics to support the U.S. Sixth Fleet, whose Area of Responsibility (AoR) to this day includes the Mediterranean, the North Sea, the Eastern Atlantic Ocean, the Black Sea, the Baltic Sea, and the waters surrounding the African continent.[47]

"Hey Mak – Top wants to talk to you."

---

[46] For a comprehensive history of VQ-2, see: http://www.globalsecurity.org/military/agency/navy/vq-2.htm Retrieved Dec. 2, 2016. // For an article regarding the disestablishment of VQ-2 on Aug 31 2012, and its integration into VQ-1, see: http://www.whidbeynewstimes.com/news/152689545.html Retrieved Dec. 2, 2016.

[47] http://www.c6f.navy.mil/about/area-responsibility Retrieved Dec. 2, 2016. The U.S. Sixth Fleet AOR covers all of Russia and Europe, and nearly the entire continent of Africa. It encompasses 105 countries with a combined population of more than 1 billion people and includes a landmass of more than 14 million square miles. It covers more than 20 million square nautical miles of ocean, touches three continents and encompasses more than 67 percent of the Earth's coastline, 30 percent of its landmass, and nearly 40 percent of the world's population.

"Top" is Marine Corps speak for Master Sergeant / E-8, and in this case Top Barker was following up on the earful that Gunny Cranston had spilled some days earlier about me wanting to transfer out of TEE-BOW.

"Reporting as ordered, Top"

"Petty Officer Mak, Gunny tells me you want out. He says you're squared away, nearing promotion, so I checked with 302 Division, and they're looking for Russian linguists to volunteer for submarine or airborne programs. Might get you where you want to go. Interested?" Top Barker leaned back in his chair, urging a quick response.

"Sure, Top – I'm here to work my share. Airborne or submarine duty is fine by me."

"Ok, then after duty report to 302's Master Chief Ulrich, and tell him I sent you."

After duty I went over to 302 "Direct Support" (or DIRSUP) Division, and told CTICM *(CTI – Master Chief / E-9)* Ulrich about my conversation with Top. There was a mystique element about the DIRSUP shop, an element that separated its sailors and marines from Rota's shore-based personnel. Much of the mystique had to do with danger[48],

---

[48] For a VQ-2 "In Memoriam" narrative, retrieved Dec. 2, 2016 see: http://www.portlyautey.com/ECM-2.htm / One of the most pertinent mishaps: **VQ-2 Bureau Number 144852, Ranger-2 CREWNAMES: KILLED: (3)** LCDR Roger Blaine Thrasher, Pilot, LT Thomas L. Walls, Navigator, AO1 Floyd Russ Bond, Passenger who was on his first A-3 Skywarrior flight, heading to Rota Spain to reenlist. **SURVIVED: (1)** AMS2 Sam "Rosey" Rozier Plane Captain. Note: Rosier continued his naval career retiring as an E-9 // On 02-26-1970 the EA-3B Skywarrior was launching from the deck of the USS Roosevelt CVA-42. The catapult system malfunctioned in mid-stroke, resulting in the Skywarrior "dribbling" off the bow. It was later discovered that the 3rd class that fired the catapult got a faulty cat shot indication, so he immediately pushed the retract button in trying to stop the launch. This in turn sheared a shear pin, causing the aircraft to go off the cat at the speed which they had reached prior to the 3rd class hitting the retract button. A faulty wire was determined to have caused the light to flash. (Note: Because of the circumstances surrounding this mishap, from

travel, and hardship, all of which equated to **adventure** in the mind of any young red-blooded patriot.

"Let me explain what you're getting into, sailor. First you'll have to qualify as a crewman, which means that on top of physical fitness and psychological suitability you'll have to know the platform, whether aircraft or submarine, like the back of your hand for employment and survivability. Then you'll have to certify in "target" knowledge and COMINT (*COMmunications INTelligence*) report-writing, all while maintaining your Russian language proficiency".

"Master Chief – as long as I can side-saddle with the best, I see no problem jumping any hurdles and certifying before the deadline. I'm just not sure whether to volunteer for subsurface or air."

"Well, you're married, so I suggest you talk it over before committing because the last thing we want is to get you through the training, and for you to then "de-vol" *(de-volunteer)* over family issues."

"I'll talk to my wife, Master Chief."

---

that point on all cat shot operators were required to immediately raise their hands above their heads when the cat is shot). As Sam Rozier recollects - Once the EA-3B got off the deck, the pilot was pulling the yoke up as far as possible trying to gain altitude, he turned and asked AMS2 Rozier what was going on? And he yelled "**COLD CAT!**" The aircraft travelled to only about 300 yards in front of the bow of the ship and began to cart wheel and break up, the nose of the aircraft broke off. AMS2 Rozier went under the water in his seat, he thought still in the cockpit, after 2 minutes of going down he disconnected his seat harness and shot up fast to the surface. Once on the surface he saw the aircraft and 3 helmets floating, at that time the USS Roosevelt ran over him and the A-3. AMS2 Rozier was swept down the side of the carrier, getting cut up from the big barnacles, breaking both of his legs, nose, and deeply gashing his head. Upon approaching the fan tail he was worried about the aircraft carrier's revving 13' prop blades, and thought it was all over, but fortunately a ship's wake swept him out to sea, away from the carrier. He was later picked up by the USS Adams and treated and flown to Naples for hospital care. "It's amazing he survived". A similarly more serious catastrophe is recorded of the loss of an entire Skywarrior crew off NIMITZ: https://www.youtube.com/watch?v=T39B-R_yXbM

“One last thing” CTICM Ulrich meant business – he lowered his tone to deep gravel.

“You fail... you SAIL!”

The admonition had a different meaning to it now. As a CTI3 *sailing* meant deploying on surface ships. Nothing wrong with that, albeit no volunteer pay kick. Furthermore, there was a prestige factor in the fast-moving brown-shoe, flight-suit and leather jacket aviation world that the black-shoe “ship-driver” Navy didn’t boast.

On the way home I realized that going “DIRSUP” was the real thing, the ticket to getting up-close and personal with the Soviet Bear, and the prospect of open-ocean adventure had already snared me. There was no going back and settling into a shore duty assignment. The sunset looked amazingly red. “Red skies at night – sailor’s delight! Red skies in the morning – sailor take warning!

In my heart of hearts I knew that submarine duty, with deployments of up to six months, would not fit family life. As a dad looking forward to a family of four, or perhaps even five before my tour in Rota was up, I knew that submarine duty would be a very hard sell with Victoria. Airborne duty, on the other hand, was usually limited to deployments that lasted anywhere from a week to two months. Victoria and I talked things over, and agreed that airborne duty was in our cards.

At the crack of dawn the next morning I had one of the most important conversations of my nascent naval career. After I told him that I was volunteering for airborne cryptologic duty, CTICM Ulrich called Petty Officer Second Class Sparks into the office.

“Sparkie – get Mak over to VQ-2. I want him to meet with our T-birds” (*CTT’s, Special Signals Cryptologic Techs*).

“Aye Aye, Master Chief.” Sparks was sharp, short, and strack, and unlike the ground-pounder sailors I had been working with, wore creases in his dungaree short-sleeve shirt. The creases split the chest pockets in half and made his sleeves stand at attention. We climbed into the duty van, and drove along the base perimeter road from the Wullenweber to VQ-2's flight hangar and Quarterdeck building across the runway, about three miles away.

Pulling into a Visitor parking spot in front of the Quarterdeck building, my eyes feasted on a neat line of four EP-3E Aries 4-engine turboprop aircraft on the tarmac, and further away a couple of EA-3B twin-jet Skywarriors parked at an angle, as if to signal that they were on "ready alert". The NSGD Wullenweber CDAA loomed large, like an Elephant Cage (*its well-worn term of endearment*) about two miles away, on the other side of the runway. Above the building portal hailed a sign with the Squadron Motto:

"Eternal Vigilance is the Price of Liberty"[49]

Inspirational. Above the Squadron Motto perched its Logo: The SANDEMAN, a black silhouette of a caped Andalusian "Don" on Spanish colors' red and yellow backdrop. The Sandeman was presumably extending a Sherry glass topped with fine Port wine (or was he pointing a pistol your way?).[50] Sparks and I checked in at the Quarterdeck.

---

[49] Traditionally, the most famous use of "Eternal vigilance is the price of liberty" that's included in books of quotations is from a speech made by the American Abolitionist and liberal activist **Wendell Phillips** on **January 28, 1852**. Speaking to members of the Massachusetts Anti-Slavery Society that day, Phillips said: ***"Eternal vigilance is the price of liberty; power is ever stealing from the many to the few.*** *The manna of popular liberty must be gathered each day or it is rotten. The living sap of today outgrows the dead rind of yesterday. The hand entrusted with power becomes, either from human depravity or esprit de corps, the necessary enemy of the people. Only by continued oversight can the democrat in office be prevented from hardening into a despot; only by unintermitted agitation can a people be sufficiently awake to principle not to let liberty be smothered in material prosperity."*

[50] Origin of the Sandeman figure and mystique: In actuality the Sandeman trademark represented Mr. George Sandeman, founder of the "Bodegas Sandeman" in Jerez de la Frontera, a town located about 25 miles Northeast of

"You'll find the T-birds behind the Green Door" the Petty Officer of the Day said as he hung up the phone from contacting CTT1 Wachdorf about his visitors.

The *Green Door* referred to a Top Secret, Sensitive Compartmented Information (TS-SCI) vault door, on the second floor of the Squadron building. It was in fact painted green, and centered below a peep-hole it had a 12-inch symbol of a bald eagle holding a crossed lightning bolt and a quill in its claws. The door was to be opened with five clicks on a cipher lock, or by ringing the visitor buzzer, which is what Sparks did.

"Third Class Makfinsky – Master Chief Ulrich wanted you to hear first-hand what airborne cryptology is all about, so take a seat and listen up." Wachdorf took station by a dry-marker board and adroitly drew a straight line, segmented it, and produced a Venn diagram on top of it.

"Here's the RF spectrum, and the chunk of it that you'll be working is primarily VHF (very high frequency), where Soviet military tactical communications take place. We CTT's, on the other hand, can

---

Rota. The Sandeman figure was conjured up by a Scottish artist who was commissioned by George Sandeman to design a logo for his wine business. Wildly successful, over the years the Sandeman figure became emblematic of the fine port wines and sherry that this Jerez region produced. Sandeman also produces fine Portuguese Port wines. The VQ-2 squadron adopted the Sandeman logo in the 1960's with the arrival of its first EP-3 Aries aircraft.

move all over the spectrum, starting with ELF (*extremely low – 3-30Hz*), which is being researched for deep submarine communications, to VLF (*very low – 3-30Khz*) – currently used for sub comms, to HF (*high – 3-30Mhz*), VHF (*very high – 30-300Mhz*) and UHF (*ultra high – 300-3000Mhz*) -where most of the tactical military comms reside, to all the way up into EHF (*extremely high – 30-300Ghz*) where some Satellite-to-Satellite and Satellite-to-Ground Station microwave relay signals are found. Typically your VHF and UHF bands are merely line-of-sight signals, meaning that their ground footprint only covers a 30-mile diameter. The good thing about being on an aircraft to collect and exploit RF signals is that by increasing our antenna height to 25,000 or 30,000 feet we can achieve a radio horizon of up to 412 kilometers (*square root of the antenna height in mtrs. x 4.124*) for a signal coming from an antenna at sea level. If you raise that transmitting antenna to 40,000 feet – assume the altitude at which an incoming TU-16 Badger bomber aircraft is cruising at, you could theoretically intercept that signal from a distance of 888 kilometers (*the sum of both of these radio horizons*). Now that's something that a Battle Group Commander is really interested in!"

The more that CTT1 Wachdorf rambled on, the more excited and enthusiastic he became.

"Enough chalk talk. Let's get a listen to some of our recordings, and I'll show you what signal exploitation is all about."

We headed into a lab with several banks of wide-band recorders installed on equipment racks. Not all racks were configured the same, but most had some variant of a frequency spectrum analyzer, as well as a time code generator or signal repeater, and an oscilloscope (O-scope). Some had a signal generator, and others housed that Sonographic device, already familiar to me. Wachdorf went over to a storage rack containing archived wide-band tapes, and drew three of them. He proceeded to put a tape onto a recorder/player amid one of the signals analysis work-stations.

"This will make your socks go up and down!" "It's a signal caught off of a Soviet Navy KA-25A Hormone Anti-Sub-Warfare helo that was engaged in ASW operations in the East-Med, north of Egypt. We found out that this burst signal contained data from his dipping-sonar sensor. We determined from COMINT provided by our CTI Russian linguist that the Hormones did a live prosecution of a UK-Royal Navy sub that was operating north of the Suez Canal. Talk about kick-ass! The signal contains tracking info, which goes to an ASW operations center (ASWOC) on its mother ship, the Helicopter Carrier "Moskva".

"Furthermore... (now Wachdorf's blood pressure was peaking) the *pair* Hormone can position itself at peak operating altitude, about 5,500 meters, or 18,000 feet to serve as a communications relay platform."

"So?"

"Like I told you, from that 5,500 meter altitude it can relay the target submarine contact datum via burst signal to ALL the other strike platforms located within a radius of FIVE HUNDRED Klicks (*kilometers*), depending on its transmitter power."

"This is kick-ass stuff, Petty Officer Mak. One can assume that the position of the target sub is then relayed back to command authorities on the 5th-E Flagship, who then confer with Moscow on whether or not to prosecute the target. If they decide to strike, they can do so with multiple launches from all of the capable killer ships within that comms bubble."

"Here – take the headsets."

While he tweaked the dials on a spectrum analyzer – yielding green worms racing across a scope, I put headsets on and... sure enough – here came the kind of sound that I had learned as a TEE-BOW operator: a burst data transmission that sounded like a fingernail scratching sandpaper. There were several scratches in succession, a long pause, followed by a steady pattern of scratches.

"These burst signals contain range, azimuth, depth, water temperature, and towards the end, the projected course and speed of the target sub."

"There's also control info that tags the transmission to this particular helo platform and OKA-2 dipping sonar rig."

Needless to say, after that techno-pep-talk I signed my volunteer paperwork to join the ranks of 302 Division's Airborne Spooks. Less obvious is the fact that I would remember that conversation with Wachdorf vividly when years later I was on an EP-3E Aries mission covering Helicopter Carrier *"Moskva"* operations in the CentMed (*central Mediterranean*). As I was exploiting the COMINT associated with several KA-25A Hormones engaged in submarine prosecution using dipping sonar, the target VHF transmissions from the helos, (who always operated in pairs for leap-frogging), took an interesting turn. The pilots were in conversation with their ASWOC on the Carrier.

"This is Raduga-22. My probe indicates live contact; steady at azimuth 96, range 12 cables (*a cable is one tenth of a nautical mile or approximately 100 fathoms*). Target course 210, speed 8 knots, depth 60 meters. Request permission to drop the DEVICE."

"Raduga-22, this is Kamen-1. Close on the contact, and when within 5 cables permission granted (*Ru: Vam Dobro*) to release the DEVICE."

The two Hormones continued to prosecute the submerged target.

"This is Raduga-21. My target update: steady at azimuth 91, range 8 cables, target course 210, speed 8 knots".

"This is Raduga-22. I am now closing (silence) dipping sensor (silence). Live Contact, steady at azimuth 90, range 4 cables. Target course 210, speed 8 knots, depth 70 meters"

Within a matter of minutes, Raduga-22 was back on the air.

"Releasing DEVICE... 4... 3... 2... 1... RELEASED NOW, at 16:59:36 Moscow time".

"On reverse course, climbing to 5000 meters."

"This is Kamen-1. Roger, on return course, climbing to 5000 meters. Report when at five thousand."

We on Ranger-22 (*all of the VQ-2 aircraft were "Rangers"*) were walking on clouds! This was the first time that a KA-25A had been caught red-handed delivering (simulated, of course, but live as can be) what we surmised to be a NUCLEAR a depth charge. We knew for a fact that the Hormone-Alfas carried conventional depth charges. Analysts were convinced, however, that conventional high-explosive (HE) charges would be released only by the partner Hormone after it had leap-frogged over the most recent HOT datum (i.e. target sub location). By releasing the depth charge from an approximate distance of 4 cables (four tenths of a nautical mile) in this exercise, however, the weapon HAD TO BE a NUCLEAR depth charge (possibly an RA-115 tactical nuclear warhead), and these Hormones were said to carry 5 Kiloton nuclear depth charges. Upon detonation of this type nuclear depth charge the target sub would be immediately torn apart by the blast's tremendous underwater shock wave.[51] As a matter of fact, any sub within a 1-kilometer radius of the blast would be out of commission not only from the shock wave, but from the effects of the severe electro-magnetic pulse (EMP). Submariners do live a high-stakes existence, knowing that weapons such as this were

---

[51] Underwater Nuclear Tests "Operation Hardtack: Underwater Tests" part 1-2 1958 DASA: https://www.youtube.com/watch?v=HWqh1qiqx8s. Retrieved Dec. 2, 2016.

designed for the sole purpose of sending them and their "Sherwood Forest" (*refers to the sub's 130-ft. Trident SSBN missile compartment containing 16 Tridents*) to Davy Jones' Locker.

Much of the post-mission reporting discussions centered around whether or not the KA-25 helos would survive such a nuclear blast, and we concluded that the pilots were goners, heroes of the "fatherland" (Ru: *Geroi otechestva*). In retrospect, however, it's possible that by climbing fast enough to get above the explosion's 6,000 foot plume[52] the helos

---

[52] Le Méhauté, Bernard; Wang, Shen (1995). Water waves generated by underwater explosion. Advanced Series on Ocean Engineering. 10. World Scientific Publishing. ISBN 981-02-2083-9. **Shallow underwater explosion:** The 1946 Baker test, just after the chimney had broken through the cloud and the crack had formed on the water's surface. The Baker nuclear test at Bikini Atoll in July 1946 was a shallow underwater explosion, part of Operation Crossroads. A 20 kiloton warhead was detonated in a lagoon which was approximately 200 ft. (61 m) deep. The first effect was illumination of the water because of the underwater fireball. A rapidly expanding gas bubble created a shock wave that caused an expanding ring of apparently dark water at the surface, called the slick, followed by an expanding ring of apparently white water, called the crack. A mound of water and spray, called the spray dome, formed at the water's surface which became more columnar as it rose. When the rising gas bubble broke the

would have been able to recover aboard the *Moskva* to await the inevitable Sixth Fleet counterattack in which the entire ship's company would become *Geroi otechestva*.

Skywarrior

As Nicky new-guy, my volunteer papers freshly signed, I was destined to first qualify as an EA-3B Skywarrior[53] cryptologic language operator. Only after pinning on EA-3B aircrew wings would I be considered for the much less arduous duty onboard EP-3E Aries aircraft.

What was so tough about Skywarrior[54] aviation, you ask? Well, for one, it included take-offs and landings from an aircraft carrier, no small feat for the heaviest (37 metric tons when fully loaded), largest (76.5 ft. long X 72.5 ft. wingspan – its wings and tail had to be folded to fit it into the hangar deck) jet aircraft to, at that time, operate from carriers.

---

surface, it created a shock wave in the air as well. Water vapor in the air condensed as a result of a Prandtl-Glauert singularity, making a spherical cloud that marked the location of the shock wave. Water filling the cavity formed by the bubble caused a hollow column of water, called the *chimney* or *plume*, to rise 6,000 ft. (1,800 m) in the air and break through the top of the cloud. A series of surface waves moved outwards from the center. The first wave was about 94 ft. (29 m) high at 1,000 ft. (300 m) from the center. Other waves followed, and at further distances some of these were higher than the first wave. For example, at 22,000 ft. (6,700 m) from the center, the ninth wave was the highest at 6 ft. (1.8 m). Gravity caused the column to fall to the surface and caused a cloud of mist to move outwards rapidly from the base of the column, called the *base surge*. The ultimate size of the base surge was 3.5 mi (5.6 km) in diameter and 1,800 ft. (550 m) high. The base surge rose from the surface and merged with other products of the explosion, to form clouds.

[53] The production designator is A3D-2Q. For some aircraft design details, see: http://tailspintopics.blogspot.com/2010/09/mighty-skywarrior.html. Retrieved Dec. 2, 2016.

[54] http://www.a3skywarrior.com/ready-room/a-3-aircraft-specs.html Contains a wealth of information regarding the A-3 Skywarrior in all of its variants, performance, and history. Retrieved Dec. 2, 2016.

Because of its size and weight, the Skywarrior's derogatory nick-name was "The Whale", and aircraft carrier flight-deck crews were always eager to describe the "Whale dance" or "Whale bounce" they witnessed when Skywarriors performed arresting gear recoveries. Its size and weight posed risks on carrier landings, when its tail-hook that was anchored to the aircraft's keel would stress under the entire brunt of the massive bird's 30-to-37 tons, as well as on catapult launches, which challenged the tremendous horsepower of these catapult mechanisms.

A series of deadly accidents looked us volunteers squarely in the eye[55]. The most tragic 1970's VQ-2 Skywarrior accident on record occurred on February 26th, 1970, when Skywarrior Bureau Number 144851 was lost at sea during take-off from USS Franklin D. Roosevelt (CVA-42). The aircraft failed to gain flying speed on its catapult-launch and crashed into water directly ahead of the ship. Four crew members were lost, while the plane captain miraculously survived and was picked up by the plane guard helo. This accident originated from what carrier

---

[55] http://www.a3skywarrior.com/personnel/memorials/a-3-accidents-by-date.html. Retrieved Dec. 2, 2016.

aviators call "a cold cat", where the catapult either malfunctions or is triggered without enough power to give the aircraft sufficient momentum for take-off airspeed at the point where the aircraft clears the carrier deck.

That tragedy had been followed by a fortunately successful "ditching" (aircraft landing in water) maneuver when on March 8th 1974 Skywarrior Bureau Number 142257 was successfully ditched at sea while attempting to recover aboard the USS America (CVA). The arresting cable #1 (a.k.a. cross-deck pendant #1) held onto the Skywarrior for approximately 190 feet, but its purchase cable (located below decks) suddenly gave way, releasing the aircraft. The pilot cunningly applied full power as the aircraft went past the left angled deck, causing it to impact the relatively calm ocean in a nose high attitude. All crewmen escaped through the upper hatch while the bird floated on the surface of the Mediterranean for about 5 minutes. They were all rescued by the America's search and rescue (SAR) helo (see pictures taken from the ship, and from the SAR helo).[56]

Then again calamity hit VQ-2 on July 9th, 1974 when its trainer TA-3B crashed after taking off in Naples Italy on its way back to Rota. Bureau Number 144863 rolled to a 30 degree climbing turn when its nose fell as the angle of bank continued to increase. The aircraft became fully inverted until impact 1.5 nautical miles from the Naples runway. The entire aircrew perished.

TOL's and FCLP's

The only way for a Cryptologic Technician to start his Skywarrior aircrew qualification process was to hitch rides on squadron aircraft engaged in Take-off and Landing (TOL) and Field Carrier Landing Practice (FCLP) training missions for new pilots and navigators. So every morning,

---

[56] http://www.portlyautey.com/ECM-2.htm - entry under "AE-3B BUNO 142257 Ra-11". Retrieved Dec. 2, 2016.

at about 6:30 am, several other trainees and I would assemble at the Squadron maintenance hangar to report for training flights.

"Hey – you three spooks, get your flight gear on. We're taking a van out to Ranger-12. We're only going one cycle (*one launch-and-recovery cycle of aircraft carrier air-wing operations usually equates to 90 minutes, but can be extended or shortened depending on number of launches and recoveries called for in the Air Tasking Order - ATO*), so we should be back on the deck by 11."

And we spooks would suit up, sling our helmet bag with O2 mask and our EA-3B NATOPS (Navy Air Training Operating Procedures Standardization) manual over our shoulder, and clamber up the airplane's integrated ladder into the cold airframe. After securing the hatches, firing up the engines, and completing the aircraft status checklist, the mission commander (usually the pilot), would do an ICS (Internal Communications System) check with the back-end crew, to include us spook trainees.

"PO3 Makfinsky ready for take-off" I would answer back.

"We're going to have a good training sortie, Navigator!" Lt. Jenkins quipped to his navigator trainee. "These spooks look alive today, bright-eyed and bushy-tailed, ready to pull some G's."

Pulling some G's was short for banking sharp turns at high air speed, where centrifugal force would increase the "Gravities" acting against your body. Those were positive G's. Negative G's took regular gravitational pull off your body when the plane drops out of the sky or is in violent stalls.

Take-off is a cakewalk, but after gaining altitude, Skywarrior pilots turn into wanna-be fighter jocks (not all of them, to be sure, but back in the 70's many), and jerk that yoke around to let their steed know who is master. Their goal was to "train" greenhorn land-lubbers on the vicissitudes of airborne physiology by subjecting them to enough positive and, on extremely rare occasion, negative G's so as to put their breakfast (or lunch) to test. And test the greenhorns they did, all the while asking questions about NATOPS procedures over the ICS.

"Spooks – answer up – What are your ditching procedures?" came another "Jeopardy"-like question from the pilot over the ICS.

"Strap-in, gloves on, O2 on, visor down, loose gear stowed, brace for impact."

"And who opens the upper hatch after ditching?" There were three hatches – the main one on the aircraft's belly – used for ingress and egress, positioned just ahead of the antenna cowling (referred to as the "canoe"), a side hatch over the right (*starboard*) wing in the aft aircrew compartment, and an overhead hatch at the top of the cockpit, alongside the "Plane Captain's" posit. If the Skywarrior somehow ditched in an inverted attitude, and survived, the belly hatch would be the upper hatch, and would have to be opened by activating the "explosive" treadle to overcome its weight.

"The Plane Captain" came the answer. At this, "Sarge", the Plane Captain looked aft making eye contact with us greenhorn spooks, smiled, and issued a patronizing thumb up for the correct answer. He was quicker out of the holster with thumbs down for wrong answers. Sarge also flipped the occasional "bird" for sorely wrong answers, and sometimes just for the hell of it.

The Plane Captain (PC) was the senior Squadron enlisted aircrewman onboard, usually an Aviation Machinist Mate –abbv. "AD", or an Aviation Electrician's Mate, abbv. "AE", although in some cases PRs (*Aircrew Survival Equipmentmen – better known as Para-Riggers, hence abbv. "PR"*) had made the effort and gone through the rigor of qualifying as Skywarrior PCs"[57]. He was in charge of the condition and maintenance

---

[57] https://www.thebalance.com/navy-enlisted-rating-job-descriptions-3345783. Retrieved Dec. 2, 2016.

of the aircraft and its avionics systems. He usually sat a posit (Poz-3, the pilot being Poz-1, and the Navigator Poz-2) located directly behind the pilot, facing aft. The PC's that I knew could *NEVER* be accused of being *Politically Correct*, and were wont to harass us spooks without respite. Their nagging was constant:

"PO3 Beerman – go clean, and I mean polish, the windshields... no bugs, no scratches, no marks!" or,

"CTI Wienner – after landing pick up the drogue chute, pack it in its bag and bring it back to the Squadron Para-rigging shop." And,

"We need some grub. Seaman Schmuckatelli – get down to the mess deck and bring back our box lunches... and do it pronto... we're about to launch in 10 mikes (*aviator-ese for minute*)!"

When I became a veteran Skywarrior aircrewman, an aviation-hardened "cryppie" as it were (i.e. I could *GET* some respect!), I would often take up station at "Poz-3", with a commanding view of the sky, and immediate access to the pilot and navigator. Sitting Poz-3 also meant that I had to assume the NATOPS duties assigned. After about 100 hours of Skywarrior flight time I knew all of the aft-crew stations, from Poz-3 to Poz-7. I was never able to sit Posits One or Two but had, on occasion, when *Charlie* was flying the aircraft (*Charlie was also known as "Otto Pilot"*), been invited, under supervision, to become *aircraft commander for a mike*.

Within a couple months of TOLs and FCLPs, which included three "arresting gear-assisted landings", and 50 hours of EA-3B flight time, I took my NATOPS certification exam, and passed it to get the hell out of the TOL phase for once and for all. Seats on those stinkin' TOLs would remain open for greenhorns and low-intensity CT's who needed flight time to stay qual'ed for lack of *REAL* operational work. I was now ready for the next phase: Aircraft Carrier Operations.

The Moving Island: USS CV(N)

There is nothing more majestic on the surface of the ocean than an aircraft carrier with its full air wing complement. These vessels are manned by up to 6,000 personnel carrying up to 90 fixed wing and rotary wing aircraft at peak capacity. Veritable combat cities with 24/7 Air Traffic Control to simultaneously conduct airborne shore strikes, anti-submarine operations, air defense, air-to-air combat, open-ocean anti-ship strikes and all-purpose, all-weather electronic countermeasures/support measures -(the Skywarrior was an open-ocean, all-weather Electronic Support Measures – ESM platform)- operations, they deliver unparalleled flexibility in the pursuit of global diplomacy.

In the '70s there were two types of aircraft carriers: conventionally powered "CV" (many of these re-designated from CVA, or Attack carriers, to multipurpose carriers when Anti-Submarine Warfare capability was added), and NIMITZ-class nuclear powered "CVN", with all the brain and brawn that forms the basis a Carrier Battle Group (CVBG).

It was the Second Fleet headquartered in Norfolk, Virginia that chop'ed its CVs and CVNs over to Sixth Fleet for Mediterranean operations, although when circumstances dictated, a carrier assigned to Seventh Fleet (Pacific and Indian Ocean Area of Responsibility, headquartered in Yokosuka, Japan) could transit the Suez Canal into the Med and take on Sixth Fleet duties. To the best of my recollection the list of US CV's that operated under Sixth Fleet's control in the 1970's were: USS Franklin D. Roosevelt CV-42 *"Rosie"*; USS Forrestal CV-59 *"FID"*, also *"Forest Fire"*; USS Saratoga CV-60 *"Sarah"*; USS Independence CV-62 *"Indy"*; USS Enterprise CV-65 "Big-E"; USS America CV-66 *"Big-A"*; USS Kennedy CV-67 *"JFK"*; USS Nimitz CVN-68 *"Old Salt"*; and USS Eisenhower CVN-69 *"Ike"*. During my tenure as a NSGD Rota Airborne Division CTI-Russian (also qualified in Spanish and French), I had the pleasure of serving onboard, operating and flying from all of these fine combatant capital ships, with the exception of the USS Forrestal (so, my friend, I can't be blamed for starting any of those fires, while I could possibly be accused of being the Phantom Shitter onboard the Indy, were it not for my fine, upstanding character ~).

For all I knew, VQ-2 Detachments onboard Sixth Fleet Carriers were established well in advance of the Carrier chopping over to Sixth Fleet operational control (OPCON) for Mediterranean operations. Perhaps these Dets were established in Norfolk, during pre-deployment work-ups. What I did know is that I only flew out to, and operated from Carriers in the Med that already had fully established VQ-2 maintenance detachments.

Because VQ-2 was an Electronic Support Measures (ESM)[58] Squadron, its Ready-Room (*i.e. the place where air combat mission pre-flight and post-flight briefings and hot wash-ups were conducted*) was shared with other squadrons that had a larger presence onboard, squadrons such as E-2C Hawkeye Early Warning and Air Controller (EWAC) squadrons, or the EA-6B Prowler Electronic Warfare (EW) squadrons. The huddles conducted in CV ready rooms are brisk, to the point, and set the pace for flights in which combatants experience seconds of sheer joy/terror, and hours of excruciating boredom.

"So that's the weather over the Gulf of Sidra up to and including the Maltese Archipelago today: as I said, CAVU (*Ceiling And Visibility Unlimited*) for the next four cycles. Should be good flying, good hunting." The Navigator, Lt. Sean Spikes smiled at the assembled Skywarrior crew.

---

[58] ESM is the passive detection of enemy electromagnetic (EM) emissions. The radiated energy of an emitter (e.g. radar) can be detected far beyond the range at which it returns a usable result to its user. Modern ESM can identify the actual class of the emitter, which helps identify the unit on which it is used. Passive cross-fixing between a number of units can locate a source to a reasonably small area and give some hint to direction and speed. ESM fixes are placed in three classes: Detected, Tracking and Targeted, depending on the accuracy of the fix and whether a unit's course and speed has been derived. Of course for ESM to work the adversaries must 'co-operate' by using their emitters.

"Great. Petty Officer Mak – anything to add from the SSES (*Ship's Signals Exploitation Spaces*) 06:00 Zulu (*GMT – we were operating in the Central Mediterranean, in Zulu+1 and it was now 07:15 local*) daily intel report?" The pilot, Lt. Darrel Drissel pointed his finger my way.

"Yes Sir, plenty". Without revealing methods and sources (*so as to bring the classification of the report down to General Service, or GENSER – since once in the aircraft I would then be able to provide the full Top Secret Sensitive Compartmented Information report*):

"We know that the Soviets have tasked their Anti-Submarine Warfare task force – which was recently split into two groups, one anchored off of Malta at their Valletta Island anchorage, and the other group anchored off of the Tunisian coast in the Gulf of Hammamet, to conduct open ocean ASW operations in the Central Mediterranean. Speculation is that these ASW operations, nick-named "***Syl'nyy Raskol-78***", meaning *Strong Split-78*, will kick-off today or tomorrow. Our FOSIF analysts believe that the Western group out of Hammamet represents the submarine "hunter" formation, and will be joined by Soviet IL-38 "May" ASW patrol aircraft deployed to Khadafy's "Uqba-bin-Nafi" airbase near Tripoli[59], Libya to simulate prosecuting a NATO-alliance attack submarine transiting south of Sicily. Simultaneously the Eastern group out of the Valletta Island anchorage will steam westerly to serve as the "strike" formation, and will attempt to execute coordinated anti-submarine cruise missile launches once the "hunter" group provides actionable targeting data."

---

[59] http://www.globalsecurity.org/wmd/facility/wheelus.htm Retrieved Dec. 15, 2016. Wheelus - Libya's 1950's – 60's SAC base, was renamed Okba Ben Nafi Air Base [aka Uqba bin Nafi Airfield], and went into Soviet use, an irony of the Cold War. It became a Libyan air force installation and contained the service's headquarters and a large share of its major training facilities. Both MiG fighters and Tu-22 bombers were located there. On 15 April 1986, it was bombed by the U.S. as one of the targets of Operation Eldorado Canyon.

"Is that all we can look forward to today?" Lt Drissel had a sense of humor when in the Ready Room. Once on the plane, however, that was another story, as attested to by his nickname "Downer" Drissel. I could recognize a rhetorical question when I heard one.

"FOSIF says that a Soviet Echo-II nuclear-powered cruise missile submarine transited the Strait of Gibraltar yesterday and was later detected in the Alboran basin by VP-62, heading east into the Algerian basin. FOSIF analysts think the Echo-II might be used as the "Strong Split" target of opportunity by simulating a NATO-alliance attack submarine. We should be on the lookout for its "Snoop Tray" surface search radar."

Once in the aircraft, on the catapult, with the carrier flight deck's Jet Blast Deflectors (JBDs) behind us in the up position, Lt Drissel powered the jet throttles back and saluted the Catapult Officer as our engines roared deafeningly.

"Shooter" (the catapult officer) made one final inspection looking fore and aft of the Skywarrior, and satisfied that all was ready for a clean shot, reached down, and touched the flight deck.

BOOM! WWRRREEEeeeeeeeeeeeeeeesssshhhhhhwwwwwIT!

That's my best approximation for rendering the 2-3 second sound of a Skywarrior catapult launch off the USS America[60]*. Its steam catapult can take a 45,000 lb. fighter from a standstill to 165 MPH in two seconds, assisted of course by the jet engines for 7.5 G's of acceleration. Could be that for a non-astronaut, the only other way to get this kind of a thrill is by sitting in a dragster during the split second when the green light flashes, and clutch releases with the throttle pedal-to-the-metal. Do our birds burn rubber on launch, like dragsters? Why, yes, I believe they do, but it's hard to see, with all of the steam and jet exhaust that's flowing over the JBDs.*

Forward we went slingshot'ed over the bow, into the Med's wild blue yonder. A spiral notepad with a pen threaded through the wire rushed back at me like a rock thrown by an angry Gaza Palestinian protester, and ricocheted off my closed helmet visor to impact again, somewhere behind me, as the bird gained airspeed and began its steep climb. Sitting in Poz-4 during a launch was always risky business. Any misplaced, forgotten loose items would come hurling back into the aft crew cabin, with Poz-4 at shortstop. It's no exaggeration to say that even the dust on the cockpit's dashboard and the dirt from the cockpit's floor would release and blow aft as the plane hit 7 G's of acceleration. We remained on oxygen and continued our climb to a cruising altitude of 38,000 feet.

"So what's this about "*Silly Rascal*"? Sarge, the Plane Captain, had been getting the aircraft ready during our pre-flight briefing in the Ready Room, so had missed the scoop.

"Could be one of our best SIGINT sweeps yet, Sarge. Looks like the Soviets are holding one of their largest ever ASW exercises in the CentMed. Lotta big players. Maybe even an Echo-II."

---

[60] For an overview of CVBG purpose and organization, and a detailed description of Aircraft Carrier flight ops. See U.S. Navy publication: https://www.cnatra.navy.mil/local/docs/pat-pubs/P-816.pdf. Retrieved Dec 15, 2016.

"Who's gonna snag the Snoop Tray?" Sarge knew his order-of-battle when it came to Soviet ships, subs, and aircraft, and Snoop Tray was a target-acquisition radar used by Echo-II submarines.

AW1 McCain – the crew's ELINT Specialist, remained silent in Poz-5. He would ignore Sarge until we reached altitude and got "on track." Both of these PO1's were hands down the best in their field, so a Type-A personality clash was normal, especially when McCain easily qualified as Crew Chief and he had seniority over Sarge. Rumor was that the only reason that McCain wasn't a Chief Petty Officer (CPO/E-7/AWC) was because he had missed ship's movement about a year ago. He had shacked up with a Naples Italy DoDDS (*Department of Defense Dependents Schools*) *Show-weeet* High School Phys. Ed. Teacher and, to keep him around, she must have slipped him a mickey the night before he was scheduled to catch his returning liberty launch to the USS Saratoga. Could just as well have been McCain's plan, *howsomever*. The shadow of suspicion came out at his Captain's Mast, when one of his shipmates testified that McCain was known to have a healthy dislike for the ship that he often called the "Sorry Sarah".

"We're on track" announced our Nav (*navigator*), Lt. Spikes over his ICS, which was switched to "All-Crew".

Teamwork the Airborne Way

Once you're "on track", it's all business. ELINT and COMINT operators put miles on dials to lock-onto and exploit relevant target signals. Today the searches are all computerized, but in 1978 your SIGINT Team was only as good as its Operators. All of us had to be versed in methods and sources, which is tradecraft for communications networks' frequencies and modulation types on the COMINT side, and for signal parameters on the ELINT side. While good intuition and keen situational awareness was needed, SIGINT teams relied on "technical extracts" that contained the U.S. SIGINT System's (USSS) distilled wisdom on how, when, where, and on what frequencies target entities communicated and employed RADAR sensors to conduct potentially lethal military operations. We took these essential "technical extracts" onto the plane on hardcopy water-soluble paper. They were printed in miniature type from a special "spook" typewriter, placed in document binders, and loaded into a padlocked steel box we called the "Tech Kit". All of the material in the Tech Kit was inventoried pre-flight, and post-flight, and returned to the SSES vault for safekeeping between missions.

McCain came alive: "Snoop Tray bearing two-eight-zero. Range at this time undetermined, it's a weak signal."

"Alfa Whiskey – this is Ranger-12. Echo-II Snoop Tray at our bearing zero-one-zero intercepted at 08:19 Zulu." Lt Drissel announced over the enciphered Air Combat network.

"Alfa Whiskey" was the call sign for the Air Warfare coordinator, and in this case AW was airborne on the USS America's E-2C Hawkeye, who not only operated the Early Warning APS-125 radar, which painted a 300+ nautical mile radius from a 30,000 ft. altitude, but also served as the controller for F-14 fighter aircraft going into air-to-air combat.[61] Overall

---

[61] https://www.forecastinternational.com/archive/disp_old_pdf.cfm?ARC_ID=72. Retrieved Dec. 15, 2016. The E-2C Hawkeye is the Navy's all-weather, carrier-based tactical warning and control system aircraft. It provides all-weather airborne early warning and command and control functions for the carrier battle group. Additional missions include surface surveillance coordination, strike and interceptor control, search and rescue guidance and communications relay. An integral component of the carrier air wing, the E-2C uses computerized sensors to provide early warning, threat analyses and control of counteraction against air and surface targets. The E-2C is the only carrier-based airborne early warning command post in the world, and has been the Navy's AEW command post for over 30 years. The sensor can monitor a surveillance area of six million cubic miles. Automated features reduce operator workload and optimize radar performance without the need for operator intervention. Over land, the radar can track aircraft over most terrain and ground vehicles when target density is relatively low. At sea, the APS-125/145 can track all significant naval targets (from large ships to fast patrol boats to stationary platforms) in most sea states. High-altitude flight allows the radar horizon to extend beyond that of surface ship sensors, a significant part of naval tactical planning.

Battle Group control was in the hands of the Composite Warfare Commander[62], representing Commander Sixth Fleet, in charge of a carefully woven tapestry of strike, defensive, and humanitarian/rescue capabilities. In hot war scenarios, VQ-2 aircraft would operate under both the Force Over-the-Horizon Track Coordinator (FOTC) and under Alfa Whiskey. In a hot war scenario with air-to-air combat Alfa Whiskey's hands would be full controlling F-14 Tomcat strikes against airborne targets.

WARFARE Commander or Coordinator	ABBREVIATION	BATTLEGROUP CALL SIGN
Composite Warfare Commander	CWC	AB
Surface Warfare Commander	SUWC	AS
Undersea Warfare Commander	USWC	AX
Air Warfare Commander	AWC	AW
Command & Control Warfare Commander	C2W	AQ
Strike Warfare Commander	STRIKE	AP
Air Resource Element Coordinator	AREC	AR
Helicopter Element Coordinator	HEC	AL
Submarine Element Coordinator	SEC	SEC
Force Over-the-Horizon Track Coordinator	FOTC	FOTC
Screen Coordinator	SC	AN

[62] https://fas.org/irp/doddir/navy/rfs/part03.htm Retrieved Dec. 15, 2016. The overall battlegroup commander is the Composite Warfare Commander (CWC) who acts as the central command authority for the entire battlegroup. The CWC designates subordinate warfare commanders are assigned to the CWC for air warfare (AWC), surface warfare (SUWC) undersea warfare (USWC), strike (STWC) and space and electronic warfare commander (C2W). Supporting the CWC and his warfare commanders are coordinators who manage force sensors and assets within the battlegroup. The CWC must remain cognizant of the tactical picture in all warfare areas and must be able to correlate information from external sources that develop locally. Generally, three prerequisites are necessary to adequately maintain the tactical picture: communications to disseminate information; displays to retain it; and a watch staff to understand and interpret it.

"Alfa Whiskey, this is Ranger-22. Snoop tray intercept confirmed at 08:23 Zulu. Range 145 nautical miles, bearing two-six-two." We were now partners in Surveillance, as both VQ-2 aircraft, the EA-3b Skywarrior Ranger-12 and the EP-3E Aries Ranger-22, operating out of Sigonella Naval Air Station in Sicily[63], reported to Alfa Whiskey. Truth be told, the Direction Finding (DF) system onboard the Aries bird was far superior to our Skywarrior DF.

"Alfa Whiskey, this is Ranger-12. Two IL-38 Mays are on Uqba-bin-Nafi runway 290, cleared for take-off at 08:45 Zulu." I had just relayed this to Lt. Drissel as I heard it over my headsets. The *"Strong Split 78"* COMINT game was ON!

IL-38 Mays were known as submarine killers, and somewhat resembled our own P-3C Orions attached to our VP (maritime patrol, Anti-Submarine-Warfare, and Anti-Ship-Warfare) squadrons, in that they had Magnetic Anomaly Detection (MAD) stingers, and carried sono-bouys for sensing underwater submarine activity, and depth charges and torpedoes to destroy their targets.

"This is Alfa Whiskey - Copy, Ranger-12. Break, Break – Ranger-22, confirm overhead report on IL-38 Wet-Eye intercept, at coordinates 34 degrees, 53 minutes North; 13 degrees, 5 minutes East."

"Alfa Whiskey – this is Ranger-12. I Report a Wet-Eye hit, bearing 195 degrees at 09:33 Zulu." Lt Drissel relayed what AW1 McCain was obtaining from his ELINT avionics suite. AW1 McCain had "scooped" the crew on Ranger-22, perhaps because of our higher altitude and extended

---

[63]https://www.cnic.navy.mil/regions/cnreurafswa/installations/nas_sigonella.html Retrieved Dec. 15, 2016. Among the aircraft that fly from this island base are U.S. Air Force C-130, C-17 and C-5 airlifters, KC-135 and KC-10 tankers and U.S. Navy P-3 Orions, C-2 Greyhounds and C-9B Skytrain IIs and C-40A Clippers, and Italian Air Force Breguet Br. 1150 Atlantiques. It is one of the most frequently used stops for U.S. airlift aircraft bound from the continental United States to Southwest Asia and the Indian Ocean.

range. The two IL-38's were actively painting the ocean surface ahead of them with their Wet-Eye radar, searching for surface objects as they headed north from Tripoli, into their ASW exercise area.

The Soviet 5th-E's *"Strong Split 78"* exercise was up and running, and its movements and tactical phases were being tracked in detail by the America's Battle Group surveillance resources. Meanwhile the "Big Picture" was being assembled at both FOSIF Rota, and at COMSIXTHFLEET's Flagship, the USS Albany CG-10 located in-port at Gaeta, Italy.

By the time we had exhausted our "on-track" fuel, *Strong Split* was still in the pre-positioning stage. The ASW "Hunter" group had departed Hammamet; the pair of IL-38 May ASW aircraft was enroute to their search zone; and the ASW "Strike" group had formed-up and was operating in an area about 100 nautical miles South of Malta. Both Ranger-22 and we, on Ranger-12, were flying reconnaissance tracks in between these two Soviet surface ship formations.

"Yo – Lt Drissel, Sir, I'm intercepting the Kresta-II's "Topsail" surface and air-search radar, and its "Head-Lights" target acquisition radar off of its Surface-to-Air SA-N-3 missile battery." McCain was agitated, and rightfully so. There's a possibility that we, or perhaps Ranger-22, or some other aircraft, were being used as a target of opportunity for the Kresta-II's air defense team.

"Strel'ka Odin – ya Oryol-657: Chuzhoy nakhoditsya v rajone poiska, peleng 270, distantsiya 29 kilometrov." I heard over my headsets, and informed Lt. Drissel that, in comms with his Task Element commander at 11:27 Zulu, Kresta-II hull number 657 – *Admiral Nakhimov*, was reporting an intruder aircraft operating in their search area, at bearing 270, range 29 Klicks (*kilometers*) from his position.

"Alfa Whiskey, this is Ranger-12 – ELINT hit indicates the Kresta-II has activated its SA-N 3 "Goblet" missiles, and is actively tracking a target with its Head-Lights radar. COMINT has the Kresta reporting a target aircraft at bearing 270, range 29 Klicks from its own position."

"Ranger-12,- break - Ranger-22, this is Alfa Whiskey. Be advised that we have an S-3 Viking operating west of the Kresta-II ASW Task Element. He will be taking pictures of the operation while in EMCOM (emission control), and is monitoring this AW frequency. Thanks for keeping him informed of the air threat picture." There was no doubt in our military minds that the Kresta-II, *Admiral Nakhimov,* was itself already in the crosshairs of one of Alfa Sierra's (Surface Strike Commander) many killer platforms, say, for example the USS Stump, one of our Spruance-class Destroyers, and that one of Stump's "Harpoon" anti-ship missiles had *Nakhimov's* name on it, if only in a simulated way.

It was commonplace for the U.S. and Soviet navies to snoop on each other, often to the point of triggering "Incidents at Sea". Some good examples were vividly recorded through first-hand experience by the intrepid Naval Aviator Don East, CAPT, USN (Ret.).[64] It was doubtful,

---

[64] Says CAPT Don East in his "History of the Cold War" narrative, as recorded by his daughter, Amy East: "Of the many INCSEA violation reports that I have been personally involved in, on one side or the other, one stands out in my memory as the most dangerous and bizarre. On 28 August 1976, my airborne electronic reconnaissance squadron aircraft was conducting maritime reconnaissance in the Mediterranean Sea. During this period, the US Navy frigate, Voge (a Knox Class if I remember), and other US Navy units were involved in an anti-submarine warfare (ASW) operation, tracking an unidentified submarine contact. Both the submarine target and the surface ships and aircraft involved in the evolution were exhibiting effective tactics in their dynamic cat and mouse game. Then, without any warning, the target submarine (a Soviet Navy Echo II Class SSGN) came to broached depth with about half the sail above the water near the USS Voge. At an estimated 12-15 knots, the Echo II appeared to make a purposeful direct run on the USS Voge. The Echo II struck the Voge amidships. After the collision, the submarine's sail tilted to one side, went under the frigate, and popped back to the vertical position on the other side. The collision broke the American frigate's main propulsion shaft, thus leaving it dead in the water. The Soviet Echo II continued for a ways before diving again. We later located and photographed it at the nearby Kithira anchorage. There, we noted men using welding torches on the damaged area of the sail front. Since I was not a member

however, that this USS America's S-3 Viking would stoop so low as to provoke the Soviets. A photo mission was in all likelihood expected, and perhaps even hoped for by the 5th-E Commander, whose mission was also to show maritime strength. What better way to show strength than by having your adversary observe and report on your most complex ASW and Anti-Ship exercises?

"Crew – we're heading for the Hawk Circle (*aviation slang - orbiting stack of aircraft waiting to land on the carrier*) due to low fuel, and Ranger-22 will remain on-station for another five hours to track progress."

"Damn, Skipper. We're right in the thick of this one! How about we check with TEXACO, get some fuel, and hang around for a while?" AW1 McCain was his old Maverick self again.

"Ok, I'll check. In the meantime, say we fuel ourselves. Sarge – are those box lunches still uneaten by "grub-locker" roaches?"

Sarge reluctantly pried himself from Poz-3 to make his way back to the "grub-locker", a nook at the rear of the aft crew cabin. Out came the heavy duty trash bag with the six box lunches. Sure, we were a crew of 5, and while the 6th box lunch was for good measure, or possibly for survival if we ditched at sea, we joked about how we would use the extra box lunch as a weapon. Open the bird's main hatch, toss it into space and let it find its way down the smoke stack of one of those Roosskie rust-buckets below. Given the quality of the lunches, that might even trigger an incident at sea.

"Looks like Colonel Sanders found some varmint to deep fry for us!" Sarge was a most perceptive gourmand.

---

of the INCSEA negotiating team at that time, I do not know the details of why the Soviet Navy Echo II submarine skipper make such a provocative and dangerous move. Was it an unintentional act done in the heat and competition of the ASW game, or was it an intentional act done out of the frustration of not being able to escape the effective US Navy tracking operation?"

The thing about fried chicken (possibly seagull) from an aircraft carrier mess deck is that it's about 80 percent chicken, 20 percent grease and oil – good for any scallywag's ruddy complexion. We all "rope-yarned" for an in-flight feast when Lt. Drissel heard back from TEXACO. If I haven't already mentioned it, TEXACO[65] is an aerial refueling aircraft, in this case a converted A-7 Corsair tanker.

"TEXACO is on his way – should rendezvous with us in about 30 mikes, with half a bag, enough to keep us on track for another two hours." Lt. Drissel had his O2 mask dangling from the side of his helmet, and was chomping down on a drumstick.

That KFC was tasting good (best grub for miles around!), and here it's appropriate to remind you that I was in the Zumwalt Navy, and that as a rugged, macho aviator I sported a full beard, which was generally not acceptable, given the need for an airtight oxygen mask fit. But, once again, I was in the Zumwalt Navy. Lt. Spikes once told me in the Ready Room to turn the darn beard into a goatee, but Lt. Drissel countered: "Leave the Spook alone!" Even Sarge tried to cow me into following his example of sporting an O2-mask-kosher goatee. "It's CinC-House's preference," said I, knowing that would put a stop to any needling. Sailors know that the Commander in Chief of the House (*spouse*) ALWAYS has the last word. Period.

"Crew – I'd say chow down fast because we're going to oxygen when tanking with TEXACO (*this to preclude fumes inhalation should any Jet Propulsion Fuel spill onto the Skywarrior, as it often did*), and as a matter of fact, best that you get your O2 masks turned on, and clipped to

---

[65] See https://www.tailhook.net/AVSLANG.htm for a superb lexicon of Naval Aviation slang. Retrieved Dec. 15, 2016. Some notables: **Tits Machine** - A good, righteous airplane. Current airplanes need not apply; this is a nostalgic term referring to birds gone by. By all accounts the F-8 *Crusader* was a tits machine. **Pinkie** - A landing made at twilight between the official time of sunset (or sunrise) and "real" darkness; it officially counts as a night landing, but is cheating; preferred type of "night" landing by Lieutenant Commander 0-4's and above.

one side of your helmet now". Our Nav, Lt. Spikes, was a stickler for NATOPS safety.

I clipped my O2 mask onto the side of my helmet, as did everyone, under orders. I switched the O2 delivery toggle switch to "ON", when all of a sudden...

"Whoooshh!"

Half of my beard, the left side of my face was ON FIRE! I had gone Michael Jackson as the O2 fueled a flaming grease fire on my chops! Possibly the first grease fire ever on a Navy aircraft in flight. In my hurry to chow down before TEXACO, I had slopped gull blubber on the left side of my pie hole.

I looked up in panic! A fire onboard an aircraft in flight is NO LAUGHING MATTER, yet, there he was – *SARGE,* his sides splitting from knee-slapping laughter, snorting and cackling like ol' goatee'd Satan himself.

I put out the flash fire with my NOMEX flight glove within seconds. The stink of burnt human beard was pretty gnarly, so Nav stepped up the ventilation, and the cabin temp. quickly dropped by about ten degrees.

TEXACO was calling RANGER to close into a refueling flight profile.

Skywarrior takes a Drink

The EA-3B has a nose-forward probe for drinking, unlike some birds that have a nose-lateral probe. This male probe marries up with a female drogue/basket, and once the two are locked and pressurized, fuel is transferred at a high rate. This is called the Probe-and-Drogue method of aerial refueling.[66] The process is kind of like what you go through when

---

[66] http://www.the-best-of-british.com/PlaneCrazyHeritage/airpictorial/1999/RefuellingFDC.htm. Retrieved

using a pint-sized pressurized propane bottle to refill a hand-held gas cigarette lighter. Liquid gas spurts from the nozzle until it's properly seated and squared with the lighter's valve.

Dec. 15, 2016. The probe-and-drogue refueling method employs a flexible hose that trails from the tanker aircraft. The drogue (or para-drogue), sometimes called a basket, is a fitting resembling a shuttlecock, attached at its narrow end (like the "cork" nose of a shuttlecock) with a valve to a flexible hose. The drogue stabilizes the hose in flight and provides a funnel to aid insertion of the receiver aircraft probe into the hose. The hose connects to a Hose Drum Unit (HDU). When not in use, the hose/drogue is reeled completely into the HDU. The receiver has a probe, which is a rigid, protruding or pivoted retractable arm placed on the aircraft's nose or fuselage to make the connection. Most modern versions of the probe are usually designed to be retractable, and are retracted when not in use, particularly on high speed aircraft. At the end of the probe is a valve that is closed until it mates with the drogue's forward internal receptacle, after which it opens and allows fuel to pass from tanker to receiver. The valves in the probe and drogue that are most commonly used are to a NATO standard and were originally developed by the company Flight Refueling Limited in the UK and deployed in the late 1940s and 1950s. This standardization allows drogue-equipped tanker aircraft from many nations the ability to refuel probe-equipped aircraft from other nations. The NATO standard probe system incorporates shear rivets that attach the refueling valve to the end of the probe. This is so that if a large side or vertical load develops while in contact with the drogue, the rivets shear and the fuel valve breaks off, rather than the probe or receiver aircraft suffering structural damage. A so-called "broken probe" (actually a broken fuel valve, as described above) may happen if poor flying technique is used by the receiver pilot, or in turbulence. Sometimes the valve is retained in the tanker drogue and prevents further refueling from that drogue until removed during ground maintenance.

The placement of the probe on the *Whale's* nose posed a risk to the aircrew because when fuel spilled out of the tanker aircraft's drogue basket, it splashed onto the cockpit wind-shields, and JP-5 fumes invariably made it into the cabin. Another risk was the fact that the EA-3B had several internal ducts that carried hot-bleed air from the jet engines to the power-generating turbines that our SIGINT avionics suite needed to obtain its 400Hz electrical power. Sometimes the cabin would suffer a hot-bleed air leak, which immediately pushed the pressurized cabin's temperature up.

When the bird went into drink mode, the Pilot would switch the ICS over to "Emergency" and link it to his air-to-air frequency so that the entire crew could follow his conversation with TEXACO.

"Ranger-12, this is TEXACO. Ready to extend the drogue. Tanking airspeed is 330 knots. Maintain a position 50 feet below my six o'clock." *Psshht! (TEXACO sounded like Darth Vader from Star Wars, and his mike clicks on and off were eerie, to say the least).*

"Roger, TEXACO. Taking station 50 feet below your six."

Lt Drissel was an accomplished pilot, and for him taking a drink was like Winston Churchill downing a snifter full of fine cognac… if it weren't for the extreme rate of transfer, he would probably really enjoy it!

"On station, 50 feet below your six. Airspeed 330 knots."

"Ranger-12, this is TEXACO. Releasing the drogue. Engage the basket when ready." *Psshht!*

Sounds pretty simple – two jet aircraft at 25 angels (*25,000 feet*) over the vast expanses of the Mediterranean Sea, hurling onward at 330 knots of airspeed, while setting up to transfer 5-to-6 thousand pounds of fuel… what could go wrong?

"TEXACO, this is Ranger-12. I'm probing the drogue, but NO JOY. Seems that the valve is IN-OP (*inoperable*)."

"Ranger-12, this is TEXACO. I'll reel it back in, and will pull an accelerated tight turn around you to get it unjammed. Maintain course and speed." *Psshht!*

From Poz-4 I could see TEXACO breaking away. Off to the left and out of sight went the Corsair. We maintained course, speed, and altitude.

Lt. Spikes got onto ICS: "Darrel – this second time around had better be good. We're at BINGO fuel for Sigonella right now."

"Bingo" was code for the amount of fuel required to safely recover at a full-service land base. In this case USS America's air-wing was to use Naval Air Station Sigonella Italy as the "Bingo" station.

"Yeah, well – we've been here many times before, and if we're below BINGO, the Air Boss will give us priority in the pattern." Lt. Drissel was not about to sweat it.

Just as suddenly as it had left my field of view, the Corsair re-appeared, closing in from the starboard (*right*) side.

"Ranger-12, this is TEXACO. How about we give this another try?" *Psshht!*

"Roger, TEXACO. Standing by for the basket."

Again the drogue hose let out, and the basket offered an inviting valve to mate with. We hit CAT[67] – **clear-air-turbulence…**

In an instant, what was a Sunday stroll through the park became an All-Terrain-Vehicle rally down Pike's Peak. We all know (*or suspect*) that on civilian air-lines the pilots invoke turbulence right after meals are served. They do this for crowd control. Flight attendants can then take a breather, chow down and relax. Not so on military combat missions… when turbulence hits, it ain't fake and self-induced.

"TEXACO, this is Ranger-12. CAT has me by the tail. Maintain course and speed, and I'll maneuver."

"Roger, maintaining course and speed." *Psshht!*

Again Lt. Drissel pushed forward with the probe as our bird's attitude[68], pitch, roll, and yaw was jostled about. I could clearly see how the basket now loomed over the cockpit windshields, then went long, then went short as our wings waggled. Lt. Drissel fidgeted with the throttles and the yoke, fighting the turbulence like a rodeo cowboy fights the good fight to synchronize his body on top of an angry bronco.

Finally the probe scored a bulls-eye, dead center into the basket. Fuel spewed out as probe and drogue wiggled into an airtight fit. It was as if we were on the receiving end of, *as one of my good Texan friends used to say*, "A cow pissing on a flat rock". And rain upon our windshields it did. There was so much JP-5 on the windshields that it was hard to see through them, so bad was the optical distortion.

---

[67] Clear Air Turbulence and Its Detection. Boeing Scientific Research Laboratories, Office of the Vice President — Research and Development. The Boeing Company, Seattle, Washington, August 14–16, 1968. ISBN: 978-1-4899-5615-6 (Online). Retrieved Dec. 15, 2016. Describes CAT.

[68] http://howthingsfly.si.edu/aerodynamics. Retrieved Dec. 15, 2016. Describes aircraft aerodynamics.

"Ranger-12, this is TEXACO. My fuel transfer gauge shows you're taking on fuel. Please confirm." *Psshht!*

"TEXACO, this is Ranger-12. Affirmative, we're drinking."

After what seemed like an interminable three minutes, Nav's thumb up signaled we now had half a bag, and were BINGO plus TWO... in other words, we had enough fuel to fly another cycle before we reached BINGO state once again.

"Ranger-12 – I'm done passing gas." *Psshht!*

He has passed gas, sure enough – our cabin reeked of JP-5 fumes.

"TEXACO, this is Ranger-12, thanks for the drink. I'm breaking away on low road to resume track."

"This is TEXACO – Roger, I'm climbing to high road. See you back at the Ranch, Ranger." *Psshht!*

"Alfa Whiskey, this is Ranger-12, done with TEXACO. Returning to REKKY (*short for reconnaissance*) track, course two-eight-zero, climbing to angels 39."

We spent another hour and a half working with Alfa Whiskey and Ranger-22 on surveilling this showcase of open-ocean ASW and Anti-Ship-Warfare by the Soviet VMF's (*acronym for "Voyenno-Morskoj-Flot", or seafaring-war-fleet*) Fifth Ehskadra. As we left track to recover onboard the USS America, *"Sil'nyy Raskol-78"* was still in full swing but we headed home knowing that we had contributed in full measure to FOSIF's forthcoming analysis.

Off Track – Hello, Mother

"Alfa Whiskey, this is Ranger-12 – Returning to Mother. I've been cleared by Red Crown (*Air Defense Controller*) to enter CCA (*Carrier Controlled Area – 50 NM around the ship*). I will check-in with "Strike" and merge into the stack."

During aircraft recovery phase, "Strike" and "Marshal" are the call-signs for "Pri-Fly", or USS America's Primary Flight Control located in the air traffic control tower – the island overlooking the flight deck. Pri-Fly orchestrates all launch and recovery operations within the 5 nautical mile "Carrier Control Zone" (CCZ). In the aggregate, this entire organization and infrastructure is called the Carrier Air Traffic Control Center – CATCC.

"Strike, this is Ranger-12, on return vector, course 100, or rather Mother's two-eight-zero, for 50 (*miles out*). Descending to angels 15. State 2.4 (*fuel remainder*). No alibis (*no maintenance problems or aircraft troubles*)."

"Ranger-12. This is Strike Control. Sweet-sweet. Mother is VFR, Case I. Contact Marshal."

"Marshal, this is Ranger-12. Mother's two-eight-zero for 35, angels 15, state 2.3."

"Ranger-12, this is Marshal. Case I. BRC is 325 (*Ship's heading during recovery, Base Recovery Course, the Carrier's magnetic course).* Take angels 9 and report see-me." *"See-me" is reported when the aircraft is at 10 NM and has visual contact with Mother.*

We descended to 9,000 feet, our Squadron's assigned recovery pattern altitude and reported "see-me" at 10 NM from Mother. We settled into the "stack", a counter-clockwise circle within 5 NM of Mother. Minutes went by as we bored holes in the sky. The other birds were gradually recovering, leaving the stack nicely.

The "stack" around an aircraft carrier in airplane recovery mode somewhat resembles a kettle of condors. Birds circle around to descend in orderly fashion, eagerly awaiting their chance to spiral inward toward "Mother" and to hook one of the arresting wires for yet another **mission accomplished**.

"Ranger-12, this is Marshal. You're next in the box. What's your state?"

"Marshal, this is Ranger-12. State is right at Bingo."

"Roger – standby."

And standby we did. From the looks of it, there were no other aircraft in the holding pattern, so WHAT THE HEY? A lot had transpired since we launched four cycles ago. The side of my face, for one, was really smarting from having my O2 mask pressed tight over burnt skin. We also knew that there was some detailed post-mission reporting still on the docket once we hit the deck.

"Ranger-12, this is Marshal. Deck is fouled – remain in the stack."

Lt. Drissel took it in stride, despite our BINGO state. To relieve the crew from the excessive cabin heat we were experiencing at low altitude, he had Sarge open the overhead hatch. Fresh air surged in and aired out our clammy flight suits as we waited for clearance to go into the landing pattern and "call the ball" (*when on final glide path toward the carrier deck, the aircraft pilot will "call the ball", meaning that the Improved Fresnel Lens Optical Landing System – IFLOLS is now "taken" in support of the "caller's" ongoing landing maneuver*).

"Marshal, this is TEXACO. All stragglers are in. I'm caboose." *Psshht!*

The TEXACO A-7 Corsair was often the next-to-last aircraft to recover. As long as TEXACO had fuel to give, and its own Bingo fuel to divert, there was a flying gas station overhead America for birds below Bingo to nuzzle up to. So who was the last to recover, you ask, the exclamation point at the conclusion of Flight Ops? Why, yes, it was our very own Puerto-Rican cousin "Angel", pronounced *Ahn-hell* (*a term lovingly ascribed to the rescue helicopter by any aviator who has experienced an ejection and subsequent helicopter rescue*).

"Ranger-12, this is Marshal. One of Raging Bull's (VA-37 Attack Squadron) A-7's had a nose gear collapse on recovery, and the fouled deck is now clear."

"Marshal, this is Ranger-12. Going for the Break."

The "Break" signals that the aircraft is breaking away from the Stack. It's done when the bird is about 5 NM ahead of the carrier, and the pilot will turn to port in a sharp descent for maneuvering into an approach flight path that will start from aft of the carrier.

We descended to 800 feet altitude (or *cherubs 8*), and started an inbound track from a point located three nautical miles off America's stern. Only the two Lieutenants up front could see what was going on ahead of us.

Nav made a hand gesture for Sarge to close the overhead hatch. There went our fresh air supply. Smelled like JP-5, sweat and NOMEX once again, and believe you me, what a treat that foul "aviation" air is to a hungry sailor's stomach. The Lt.'s completed the landing checklist, switched landing gear down, tail-hook down, and extended the Whale's flaps for landing.

The Whale was now on the glide path, "on the ball", so to speak, at 600 feet altitude, heading toward the angled recovery deck (80 ft. above sea level) from astern at a speed of 250 KIAS (Knots, Indicated Air Speed), which is about 288 mph. *Try parking your car in your garage while coming in at 30 mph... can't do it? Ok, now add some arresting gear and try again.*

Lt. Drissel was now on final, and "flying the ball" all the way down. So far, so good. I could see his right hand reaching for the dual-engine throttles. The pilot made minor adjustments to power, rudder and ailerons. Case-I recoveries in good weather and visibility seldom required air-to-ship communications. As a matter of fact, pilots were encouraged to maintain an EMCOM (*emission control*) profile unless an emergency occurred. By flying the ball, Lt. Drissel was getting real-time visual feedback on whether or not he was on a good glide path into the arresting cables.

Flaps were lowered further to bring the airspeed down to below 220 KIAS, in deceleration. Slats on the wings' leading edge did their own thing, occasionally clanking. Seconds turned into minutes as our jet engines carried us toward a controlled crash onto America's Bethlehem Steel flight deck.

Skywarrior raced toward America's stern, then gracefully flared nose upward, with the tailhook and tires extended like an eagle's talons reaching for steel on that calm and balmy Mediterranean evening when....

KA-BAAAMM!!! The Whale shuddered like a train derailing, and at the top of his voice Lt. Drissel yelled out:

"BOLTER – BOLTER!"

Our tail-hook had slammed down on the deck at the same time as our main wheels hit the non-skid, but there was absolutely NO braking force to slow us down. The hook had bounced over the arresting cables, Pendants 3 and 4.

I could see America's island superstructure through the tiny window in the starboard over-wing escape hatch to my right. The image was just a blur, and it flashed by in a second. Lt. Drissel pulled the throttle sticks back. Our twin jets whined like angry stallions, kicking into full military power. Over the waist deck we sped, hoping to have enough airspeed for a quick climb.

Once again we launched off the carrier deck, only this time with no catapult. Entirely on our own power Drissel climbed to resume the pattern, resolved to snag either the number One, or the number Two cable on our next pass. A "BOLTER"[69] carries no negative connotation. The physics just didn't line up. Could be that the tail-hook bounced over the number Three cable, and was still in the "up" position as the bird went over number Four, which was our last chance for an arrested recovery. The Whale doesn't do too well on number Four. Because of its hefty mass it runs Four and its purchase cable all the way out, often leaving the bird's nose wheel precariously close to the angled recovery-deck edge, with pilot and nav looking over what the Ship's crew calls the carrier's crotch, into the ocean 96 feet below. It's tough to taxi out of that spot.

"Marshal, Ranger-12. In the pattern."

"Ranger-12, Marshal. Call the ball. TEXACO's holding, in the stack. TEXACO only has a thousand pounds to give".

Landing on an aircraft carrier had once been described to me by an EA-3B Plane Captain as piercing downward from on high through the clouds, locating a tiny postage stamp floating on the deep blue sea, and then descending to the white-caps for a controlled crash. He was quick to brag that for this, he got paid about $2 per CAT (*catapult*) and TRAP (*arrested recovery*). Enlisted flight pay back then was under $100 per month. Needless to say, part of the bargain that kept us going is that we all had an agreement with Neptune: *at the end of our aviation careers our number of CATs would be equal to our number of TRAPs.*

"Pilot, this is Nav – we're way under BINGO."

"Nav – I'll pick-off One or Two this time. TEXACO's plan B."

---

[69] **Bolt, Bolter** A carrier landing attempt in which the tailhook fails to engage any of the arresting wires, requiring a "go-around," and in which the aircraft landing gear contacts the deck. Otherwise it is a "low pass." For carrier landing methodology, see - http://science.howstuffworks.com/aircraft-carrier4.htm. Retrieved on Dec. 15, 2016.

Our mood was somber. If we didn't catch the wire, our options were:

A. TEXACO for more fuel, and a thousand pounds is no bargain

B. A third attempt on fumes

C. Following a third BOLTER, TEXACO would recover, and Mother would set up a BARRICADE to catch us

D. Were we to "miss" the BARRICADE, we could DITCH the plane at sea.

Other options (*such as taking altitude for a controlled bail-out on auto-pilot*) were too far removed from reality. Even the ditching option (which I was to face much later when lost at sea with minimum fuel to reach the Azores' Lajes airfield) was one that gave the crew a minimal - "*infinitesimal*"- chance of survival.

Drissel called the ball. On glide path. All looked well. Flaps out. Nose flared up.

KA-BAAAMM!!!

VVVREEEEEeeeeeewwwwwwwwwwwwwwwwwwwwwwwwwwwwwww

At the **"K"** we're on a 30-ton Whale doing 150 knots, and within two seconds (the "w") the aircraft has reached virtual standstill, ready to taxi over to a parking area for the Blue Shirt aircraft handlers to take over. The negative G's we experience? The forward compression easily peaks at four or five, as our bodies heave against the over-the-shoulder and around-the-waist torso restraints and we bear down to keep our feet on the deck. When the arresting gear does its job, sailors are gladly subjected to the pounding.[70] Unfortunately there are times when demons take over, and the arresting gear fails, sending aircraft, and sometimes sailors, into Davy Jones' locker.[71]

---

[70] Modern carriers typically have three or four arresting cables laid across the landing area. All U.S. carriers in the Nimitz-class, along with Enterprise, have four wires, with the exception of the USS Ronald Reagan and USS George H.W. Bush, which have only three. The Gerald R. Ford-class carriers will also have three. Pilots aim for the second or third wire (depending on ship configuration) to reduce the risk of landing short. Aircraft coming in to land on a carrier are at approximately 85% of full throttle. At touchdown, the pilot advances the throttles to full power. In the F/A-18E/F Super Hornet and EA-18G Growler aircraft, the aircraft automatically reduces engine thrust to 70% once the deceleration of a successful arrestment is detected. This feature can be overridden by the pilot by selecting max afterburner (note that the EA-3 Skywarrior has no afterburners). If the aircraft fails to catch an arresting cable, a condition known as a "bolter", the aircraft has sufficient power to continue down the angled flight deck and become airborne again. Once the arresting gear stops the aircraft, the pilot brings the throttles back to idle, raises the hook and taxies clear. In addition to American CVNs (nuclear aircraft carriers), the French Charles de Gaulle, the Russian Kuznetsov, the Brazilian São Paulo, the Chinese Liaoning, as well as the Indian Vikramaditya are active or future aircraft carriers installed with arresting gear.

[71] https://www.navytimes.com/story/military/2016/04/14/terrifying-cable-snap-sent-navy-plane-plunging-off-aircraft-carrier/82936858/. Retrieved on Dec. 15, 2016. One sailor remains hospitalized and three others are in recovery after an arresting cable snapped during a carrier landing in a rare and terrifying flight deck mishap on March 18, 2016. Sailors on the carrier Dwight D. Eisenhower's flight deck saw the landing go horribly wrong in an instant that Friday afternoon. An E-2C Hawkeye snagged the metal cable stretched across the deck. But the plane didn't slow, instead careening down the runaway and plunging over the

The A-7 TEXACO Corsair and the S3-H Sea-King "Angel" recovered without a hitch. With our Skywarrior's wings folded, she was parked and chained down to deck tie-down holes.[72] The pilot and nav de-briefed in the Ready Room, while I hauled my tech kit over to SSES (again, Ship's Signals Exploitation Space – the secure compartmented information vault facility) to draft the post-mission report relating to methods and sources. Lt. Drissel, who was also the Mission Commander, would later, along with the Ship's Intel Officer, "chop" and release the report.

To Ditch or Not to Ditch

Not all Skywarrior missions were as dull as the one I just described. Perhaps one of my most adrenalin-filled adventures took place when I was part of a Skywarrior crew tasked to cover Soviet Long-Range Aviation (LRA) operations against one of our transiting Carrier Battle Groups in the Atlantic. In this case it was the USS Eisenhower CVN 69 "Ike" with the CVW-7 Air-wing onboard. Ike had departed Norfolk on 16 Jan 1979 for a rotational Med cruise to relieve the USS Saratoga CV 60 "Sarah" with Air-wing THREE onboard, and was scheduled to return to homeport in early July 1979[73]. As an interesting factoid, consider that

---

edge, according to a squadron member who has spoken to eyewitnesses and those injured. Behind it, the cable came unhooked from the port side and whipped around toward the superstructure, striking eight members of Carrier Airborne Early Warning Squadron 123. Their injuries range from cuts and bruises to a skull fracture and broken bones, according to the squadron member, who asked not to be identified amid an ongoing investigation. It had been more than 10 years since an arresting cable, a thick rope of wires that catches an aircraft's tail hook to slow it down and complete a carrier landing, has parted on a carrier deck, according to Naval Air Force Atlantic spokesman Cmdr. Mike Kafka.

[72] https://www.google.com/patents/US3298148 Aircraft Mooring technologies described here. Retrieved on Dec. 15, 2016.

[73] For a chronological listing of CVBG deployments, see: https://www.history.navy.mil/content/dam/nhhc/research/histories/naval-aviation/dictionary-of-american-naval-aviation-squadrons-volume-1/pdfs/Appendx3.pdf. Retrieved on Dec. 15, 2016.

during her Med cruise, while off the coast of Haifa, Israel, the Ike hosted Israeli Prime Minister Menachem Begin, who had recently returned from signing the Camp David Israel – Egypt Peace Treaty in March of 1979.

Typically VQ-2 would deploy reconnaissance resources to Lajes Air Force Base located in the Azores Islands – the Island of Terceira, to be precise.[74] More often than not it was the EP-3E Aries that served as the reconnaissance platform of choice, because of its longer on-station dwell time, and its more capable SIGINT suite. In terms of SIGINT aircrew alone, the Skywarrior's three ELINT workstations and two COMINT workstations paled in comparison to the Aries' six ELINT / one Special Evaluator / one Special Signals / one COMINT Evaluator / six COMINT workstations, a head (restroom), and a galley (food service nook).

In late 1978 and early 1979, however, our President Jimmy Carter was fully engaged -come hell or high water- in pushing Israel and Egypt into a peace treaty[75]. To determine whether both Israel and Egypt were playing good in the sandbox, the SIXTH Fleet stepped up all reconnaissance activity in the Eastern Mediterranean, requiring daily flight operations that alternated U.S. Air Force RC-135 Rivet Joint SIGINT

---

[74] "Lajes Field History - The U.S. Enters the Azores". Washington D.C.: 65th Air Base Wing Public Affairs/Air Force ePublishing. 6 June 2006. Retrieved Dec. 15, 2016. Describes the history and functions of Portugal's Lajes Air Base.

[75] http://www.cnn.com/2013/08/23/world/meast/camp-david-accords-fast-facts/ Retrieved Dec. 15, 2016. The peace treaty between Egypt and Israel was signed on March 26, 1979, 16 months after Egyptian president Anwar Sadat's visit to Israel in 1977 after intense negotiation. It was signed in Washington, D.C. by Egyptian president Anwar Sadat and Israeli Prime Minister Menachem Begin, and witnessed by United States president Jimmy Carter. The main features of the treaty were mutual recognition, cessation of the state of war that had existed since the 1948 Arab–Israeli War, normalization of relations and the complete withdrawal by Israel of its armed forces and civilians from the Sinai Peninsula which Israel had captured during the Six-Day War in 1967. Egypt agreed to leave the area demilitarized. The agreement also provided for the free passage of Israeli ships through the Suez Canal, and recognition of the Strait of Tiran and the Gulf of Aqaba as international waterways. The agreement notably made Egypt the first Arab state to officially recognize Israel.

birds and U.S. Navy EP-3E Aries SIGINT birds flying out of Hellenikon Air Base in Athens, Greece, or from Souda Bay Naval Air Station, Crete on Eastern Mediterranean surveillance tracks. President Carter also had a CVBG operating in the East-Med: the USS Saratoga "Sarah" CV 60 with its air wing CVW-3 was to keep a lid on any untoward military provocations by either side. That meant that VQ-2's EP-3E Aries birds were max'ed out in support of a successful peace treaty ratification.

Lajes, incidentally, is just over a thousand miles West of VQ-2's home-plate in Rota, Spain. While today navigating an open ocean one-thousand mile flight is a mere cake-walk, thanks to GPS, in 1979 aircraft navigators relied on TACAN (*TACtical Air Navigation*) radio-navigation aids interacting with their onboard Inertial Navigation Systems (INS). As we will see later, TACAN was not always reliable.

During the height of the Cold War, 60's through 80's, Lajes played host to the U.S. Navy's Naval Air Facility Lajes (NAF Lajes), a tenant activity at the air base. NAF Lajes, and its associated Tactical Support Center (TSC)/Antisubmarine Warfare Operations Center (ASWOC), supported rotational detachments of U.S. Navy P-2 Neptune and later P-3 Orion maritime patrol aircraft that would track Soviet attack, guided missile, and ballistic missile submarines in the region. VQ-2 benefited from that strong ASW presence, and used these facilities when conducting maritime reconnaissance missions against the Soviet LRA.

On a fine January morning, after bidding farewell to our families and friends the day and night before, a crew of five Navy airborne reconnaissance operators, consisting of the Pilot/Mission Commander Lt. Charles "Chuck" Jones, Navigator Lt.jg. Bradley "Blazer" Darnson, the Plane Captain (and Senior ESM-EW Operator) AW1 Russ "Rooskie" Stands, the ESM Operator AW3 Darrell Farleigh, and myself, Cryptologic Operator – CTI2-Russian "Knife" Makfinsky (I had earned my call-sign nickname through a Karaoke rendition of Bobby Darin's "Mac the Knife" at the NAS Sigonella NCO Club). Two "ground pounders" (i.e. non-flight-qualified personnel) were dead-heading on this transit flight, and were to augment the maintenance department at the NAF Lajes to support this rapid-response VQ-2 deployment.

About an hour before dawn we had formed a daisy-chain to fill Ranger-17's hell-hole with baggage, aircraft parts, and miscellaneous equipment and reconnaissance kit while a fuel truck had parked alongside to feed the bird a full bag of fuel. We were, for all intents and purposes, ready to recover at Lajes with sufficient Bingo fuel (recall that Bingo fuel means enough fuel to get to an alternate landing site) to divert to Joao Paulo Airport in the city of Ponta Delgada on the island of Sao Miguel, also in the Azores Archipelago. The Nav felt confident, however, that a full bag was not needed for this thousand nautical mile flight, so he sent the fuel truck packing after reaching three quarters full.

"Good morning, crew. Nothing special about this transit flight. The weather is good on departure, but we can expect overcast conditions or even rain showers in Lajes." AW1 Stands relayed to us after buttoning up the hell-hole. Both the pilot and the Nav were filing the flight plan at the Base Operations center in the Air Traffic Control tower. Since we were recovering in Portuguese territory, there was more to the process[76] than when we recovered on one of our own aircraft carriers.

"Rooskie – let's get this show on the road. Kick the tires, and light the fires!" Lt. Jones was raring to get airborne, and there was much to do on our arrival in Lajes, such as setting up the Detachment in time to prepare for our early launch the following day.

After a smooth take-off directly over the Atlantic, we assumed a course of 275 degrees, due West. Once we got to altitude, the ground-pounders snoozed. There would be no flight attendants, and the box lunches provided little entertainment – sleep was the best antidote to weariness and boredom. Both of them would need to hit the deck

[76]http://www.faa.gov/about/office_org/headquarters_offices/ato/service_units/systemops/fs/res_links/media/icao_flight_plan_filing.pdf. Describes procedures for filing a flight plan. Retrieved on Dec. 20, 2016.

running upon arrival at the NAF Lajes. The rest of us fired up our Poz's to ensure all was up and running to "hunt for Bear" the next day.

"Hunting for Bear" refers to gaining "indications and warning" information on the TU-142 (TU stands for *TUpolev, one of the Soviet aircraft design-and-build bureaus*), NATO designator Bear F/J Long Range Maritime Patrol aircraft.[77] These Bears had a penchant for warmer climes,

and would descend from the Arctic permafrost of their home base at Olenegorsk South of Murmansk on the Kola Peninsula, where they would take off from their 3,500 meter long runway. The Bears' sole purpose was to shadow the US CVBG as it transited the Atlantic, and to send targeting information back to all able, available, and participating Soviet

---

[77] http://www.ausairpower.net/APA-Bear.html#mozTocId258676. Retrieved on Dec. 20, 2016. The Tu-142M Bear F was Russia's premier LRMP aircraft, larger, faster and longer ranging than the P-3C Orion, and typically armed with heavier ASW weapons and a larger sono-buoy payload. It was introduced into service in 1972, and incorporated significant design changes, including integral wing tanks rather than bladders, an extended forward fuselage, an extended-chord rudder, improved undercarriage, but retaining the 'classic' glazed nose of the Bear A. It's equipped with a comprehensive ASW/ASuW avionic suite. Its large size and endurance made it unusually potent in the LRMP and ASW roles, as it could outrange and outstay smaller airframes.

strike platforms. If the CVBG was located North of an imaginary line stretching from Iceland to Scotland, the TU-142s would recover back at Olenegorsk. If, on the other hand, the CVGB was transiting South of that imaginary line, the Bear F/Js would invariably recover at their forward deployment bases at San Antonio de los Baños, and Cienfuegos, Cuba after scouring the waters off of the U.S. East coast for submarines. These occurrences were frequent, and sometimes warranted a mention in the CIA's Annual National Intelligence Estimate.[78]

As we headed West toward the Azores at a comfortable 346 KIAS (again, knots indicated air speed) cruising speed, at cruising altitude, the Sun seemed to stay with us, our steady wingman.

"Knife – any info on the Bear deployment in today's FOSIF report?"

"Negative, Lt. Jones. As soon as the birds file a flight plan, NSA will tip FOSIF off with an advisory. If the Bears don't file, however, either NSA or NSG Rota will have to get indications from breaking their Olenegorsk Air Defense Tracking data when they signal take-off."

The Bears often flew under EMCON (again – EMission CONtrol), without filing flight plans. When they flew near, or even crossed over Norwegian or United Kingdom airspace, our Allied air defense over-the-horizon radars would pick them up, and Air Defense Zone (ADZ) controllers would challenge them to Identify Friend or Foe (IFF) with their transponders. Since total EMCON meant that their transponders were in wartime mode (and would only respond to other Soviet IFF interrogations), Norway and the UK would then be forced to scramble fighters to escort the Bears away from Norwegian and UK sovereign air space. This game was constantly in play, one of the Cold War "givens". As a matter of fact, it was commonplace for military aircraft of Western

---

[78] https://www.cia.gov/library/readingroom/docs/1979-01-31b.pdf. Retrieved on Dec. 20, 2016.

Block and of Eastern Block forces to fly “inciter” missions against all kinds of Air Defense Forces in order to gain intelligence on the types of early warning sensors that the defenders used, their fighter aircraft reaction procedures and effectiveness, and their Surface-to-Air missile defense tactics and procedures.

One of my duties was to keep an eye on FOSIF reports, so I would have to check-in with the ASWOC, which served as our Intel Center, as soon as we hit the deck in Lajes, and confer with their intel analysts. No doubt I would link up with other Cryptologic Technician bubbas. My Rota homeplate had already signaled my Det’s impending arrival in Lajes to the local Intel Center Officer in Charge (OiC). They had probably already set up a daily “read board” with pertinent Olenegorsk Bear info.

Since we were on a transit flight, the Pilot and Nav had linked their mikes to Emergency/ALL Stations Intercom, and those of us in Poz’s 3 through 7 could toggle onto Emergency/ALL Stations to hear what they were chatting about. Since my COMINT gear checked out, and there was nothing for me to exploit, I toggled over to ALL Stations.

“Blazer – we’ve held course 275 for several hours now. According to OPS we have some Jet Stream effect hitting our 10 o’clock. That means that we’re experiencing a North-westerly drift. I’d expect that by now we should have made a course correction to 260 degrees or so.”

“Chuck – our Inertial Nav System (*INS*) shows us on track. We’re on course for Lajes.”

“Ok. I’ll let Charlie (*slang for auto-pilot*) take over, but keep an eye on him while I decompress.”

Lt. Jones switched the auto-pilot on, unclipped from his seat, climbed down from the cockpit to the crew cabin and used the “piss-tube” to decompress. As you might imagine, the piss-tube was a black Bakelite funnel with a trigger-operated Open/Close valve and a black hose that went through the aircraft hull to drain through an orifice on the underside of the Whale. Fail to press the trigger, and you would soon find your hand awash in golden flow. Maintenance crews would try to trick greenhorns into checking the substance along the aircraft's surface for viscosity to see whether the bird was leaking oil or hydraulic fluid.

Lt. Jones got back into his seat in the cockpit, strapped in, scratched his head before stuffing it back into his brain bucket.

"Blazer – something tells me you might best break out the sextant, and shoot the Sun."

"Well" replied Blazer "I've bounced TACAN[79] against our INS just about every 30 mikes, and have come up with good fixes. We're on course."

As an impartial observer, it seemed to me that our Nav was being defensive. His defensiveness might have been because he was a Lieutenant junior grade, and he still had to prove himself. His denial was concerning, however, and Lt. Jones wasn't buying.

"Blazer – break out the sextant, and shoot the Sun. NOW!"

Blazer reached into a compartment alongside his starboard bulkhead, and extracted a case. He broke out the sextant, and swiveled his seat around to point his face and sextant towards the Sun.

He did this three times, at ten minute intervals. His plots intersected within an "area of probable location" on his navigational air chart.[80]

---

[79] TACAN receivers operating perfectly can contain good data interspersed with bad and even "wild outlier" data. In fact, fully usable radio-navigation aids often have a few good data points interspaced with large amounts of useless measurements. To counter this, NASA applies corrective calculations by using a data rejection algorithm known as the Kalman Filter. It's doubtful that the A-3 TACAN suite incorporated Kalman Filter computers.

“Chuck – you were right. The Jet Stream must have pushed us *way* north. We haven’t made very good time at all, and need to turn to 240 degrees.”

Lt. Jones banked us to port (*left*) for a 240 heading.

“How far out are we?”

“According to this, we might be between 500 to 550 miles (*nautical miles*) out.”

“WHAAAAT? Check again, Blazer! We only have enough fuel for about 600 miles!”

While Blazer worked the sextant, Chuck tried to get on the horn with the Lajes Air Traffic Control (ATC). His attempts failed, indicating that we were still outside of the VHF comms bubble with Lajes ATC (*remember? square root of the antenna height in mtrs., which was ~ 12,000 mtrs. x 4.124 = 452 Kmtrs., or 244 nautical miles*).

“Blazer – I can’t get Lajes ATC. We must be over 250 miles (*NM*) out. If I climb I’ll extend our radio horizon, but will have to burn more fuel to get to our 44 Angels ceiling.”

Pilot and Nav conferred, and came to the conclusion that rather than climbing, it was best to turn south, to course 225 degrees. We maintained that course for a good 20 mikes. Not a peep from the in-line

---

[80] Prior to GPS, inertial navigation was the primary means of navigating across oceans. However, a great many airplanes did cross oceans before INS, with varying success. The general idea was to use a combination of dead-reckoning, radio beacons where you can find them (small islands and coastlines), the sun (with the help of a sextant) during the day, and the stars at night. Hopefully, this puts you in position after some period of time to pick up ground-based navigational aids, or to identify a coastline. This actually works pretty well if your wind estimates are good, and the margin for error is high enough. By the time inertial navigation became prevalent in the 1970's, airlines were already providing reliable transcontinental passenger flights every day.

crew, some sleeping, oblivious to our predicament, others of us in worrisome suspense monitoring the Emergency/All Stations ICS like their life depended on it. Lt. Jones continued to call Lajes ATC, and finally, **a break-through**:

"Ranger-17, this is Lajes ATC. Roger, you're in the Santa Maria Flight Information Region, but I don't have you on our radar yet. We look out to about 200 nautical miles."

"Lajes Control, this is Ranger-17. We think we're on your Radial 060, Angels 39, KIAS 346. Notify me when you paint me on your screen."

"Sextant fixes now show us about 250 nautical miles from Lajes, their radial 060." Blazer said to Chuck... only now he was *really* sweating it. His reputation was on the line, and our situation was now precarious.

"Damn, Blazer – safe landing procedure calls for enough extra fuel to divert to Joao Paulo in case of trouble at Lajes, and that's roughly 130 miles from Lajes. We're FU..ED!" *(Alas – if only the letters EL had been applied before take-off).*

"TACAN data was checking out just fine..."

Lt. Jones cut him off "Checking out just fine? We've drifted 500 miles off course! Damn, Sherlock – get us on a beeline into Lajes, and NO FU..-UPS!"

At that, Lt. Jones switched his ALL-Stations toggle off, and proceeded to holler out one good ass-chewing, the likes of which Blazer had never before been subjected to.

They talked on their Pilot-Nav ICS circuit for about five minutes. Then Lt. Jones toggled back to ALL-Stations.

"Crew – I want you to go through the procedures to ditch at sea. Lajes doesn't have us on radar, and we're below our fuel safety margin." At that, he turned around, and back-handed "Rooskie" in the arm.

"Rooskie – make sure that the entire crew rehearses ditching at sea procedures at least three times. Everyone participates, including the ground pounders."

"Aye – aye, Skipper."

Rooskie, the top-drawer Plane Captain that he was, knew the ditching procedures as if he had personally written them. By the third dress-rehearsal, the entire back-end crew knew exactly who was going to pop which emergency escape hatch (depending on our plane's final attitude at rest – which, by the way, we all thought would be nose down, racing toward the bottom of the Atlantic after doing several cartwheels upon impacting the choppy whitecaps). We also knew who was going to deploy the life-raft, who was going to take the water breaker, who was going to take the survival rations and the first aid kit, who would be in charge of activating the rescue radio-beacon, etc., etc.

What we could not venture to guess was who would be injured, maimed, or killed when the jet engines exploded from contact with the cold salt water, maybe sending shrapnel and jet-blades-sharp-as-knives through the aircraft's fuselage. Our thoughts went to our families who would only hear about our demise from a hapless messenger, be it Navy or Red Cross. This transit flight had morphed from being a Sunday stroll into becoming a veritable "SWEATEX" (*aviator slang: Sweat Exercise as opposed to Boring Exercise – BOREX*).

"Truth is, crew, that the A-3 can survive ditching at sea, and as long as the fuselage doesn't break apart, we might have a good four or five minutes to evacuate before she sinks."

Lt. Jones' words were reassuring. All we could do was to stay strong, hunkered down and conserve energy as the situation unfolded beyond our control. Lt. Charles "Chuck" Jones was God's personal envoy, sent down to Earth from On-high to guarantee our survival.

"Chuck" -the navigator came onto the ICS- "if we go low we might conserve fuel."

"Blazer – if I want your advice, I'll let you know! Fact is that we need to maintain altitude until Lajes Control sees us on their radar. If we go low, forget it. They won't be able to pick us up at 200 miles out."

The ICS went silent. It all seemed surreal. Seven perfectly able humans on an aircraft that might soon have to perform a crash-landing on a choppy open ocean, somewhere over 200 nautical miles North-east of the Azores. I've been lost on foot, lost in a car, but this was the first time ever that I was lost on a plane while out in the middle of the ocean. There is nowhere to stop and ask for directions, nowhere to pull over to regroup.

For some reason (probably to evade gloom) I thought about the time when I was on a cross-country road trip from California with my brother Steve, and Victoria, who was pregnant with Ivan, to visit my mother in Texas. It was just before Christmas 1974, and we were crossing the Arizona desert well before dawn, heading out to my Mother's in Fort Worth, Texas. Victoria was snoozing under blankets in the back seat (busy building Ivan), I was riding shotgun, and Steve was at the wheel taking his turn on our non-stop drive, speeding freely along an empty Interstate-10 East. I thought I knew the car well, and when the fuel indicator needle hit reserve, I told Steve that we still had enough gas to cross the desert to the next gas station, some 40 miles East, so we breezed past Benson, AZ, to make our next stop at the first gas station we could find, which later turned out to be at Travel Centers of America in Wilcox, AZ. I was wrong... after 20 miles of that "*igorant*" conversation the car sputtered and died. The Arizona desert is DAMN cold in late December, at 4 am. Steve hitched a ride to Wilcox, returned with a gas can, as Victoria and I huddled under blankets in the 30 degree night. *Baaaad. Bad Nav.*

If only one of those RF beams from Lajes' Air Traffic Control tower would paint the steel skin of our aircraft, we might know whether we had a realistic alternative to ditching at sea. If only...

"Ranger-17, this is Lajes Control – no sign of you yet. Request you fly zig-zag patterns, maintaining your same general heading."

Chuck knew that we would consume more fuel doing zig-zags. He decided against it, and didn't respond to Lajes Control as his brain surveyed our situation for a clear way out. He addressed the crew.

"We're going to stay on course." He said. "If you have some snacks in your pockets, eat up – you'll need the extra energy."

Out came Mars bars, Snickers, packets of Orio cookies, packets of peanut butter cheese crackers, bags of peanuts, and as we shared and proceeded to gobble down all of this high-calorie junk food, one of the ground pounders switched on his ICS mike and intoned a familiar chant:

"Eternal Father, strong to save,
Whose arm hath bound the restless wave,
Who bidd'st the mighty ocean deep
Its own appointed limits keep;
Oh, hear us when we cry to Thee,
For those in peril on the sea!"

Not a good singing voice, by any stretch of the imagination, but we all knew the Navy Hymn, and it put some wind in our sails. Nobody said anything, but we could all sense appreciation. The words resonated in our imaginations, which were now conjuring visuals of our soon to be new workstations, paddling the oversized orange survival raft.

"When I signal **PREPARE TO DITCH** I will mean it, and you've got to be ready. I want an in-line acknowledgement, NOW!"

"Poz-2 - WILCO"

"Poz-3 - WILCO"

"Poz-4 - WILCO" I acknowledged, and so did the rest of the crew. The skies below us were overcast, and the sun was waning, leaving us to take up its next station, newday dawn over Kamchatka, and soon after, the Asian continent mainland. It could be raining down below us, for all I knew.

If we survived this, Nav owed us all a paid vacation for his having been lulled into TACAN complacency. Minutes transpired as adrenalin levels surged.

"Ranger-17, this is Lajes Control – you're on our screen, Radial 040, distance 195 nautical miles."

"YEEE-HAAAWW!!" Blazer was ecstatic. If we ran out of fuel, and had to ditch, at least Lajes Control would be able to send a Search-and-Rescue chopper to pick us up out of Neptune's icy reefer.

"Lajes Control, this is Ranger-17 – I'm low on fuel, at about a 300-mile remainder, so below Bingo."

"Ranger-17 – I'm bringing you in on the Emergency Runway One-Five-Zero, Priority recovery, prepared for crash landing with fire suppressant ready. You will only have one pass. Make it count."

"Lajes Control – Roger. Course 220, Angels 39, KIAS 350. Prepared to ditch at sea if necessary."

Who knows what kind of preparations the Fire Crew was going through at Lajes? No doubt they were jumping through their grommets to put some foam on the runway in case our Whale dropped from the sky, air-mailed, so to speak.

"Crew – we're conserving fuel as best we can, but at some point I will have to decide whether to Ditch or attempt an Emergency Landing at Lajes. If we start our descent into Lajes on fumes, we run the risk of a

FLAME-OUT.[81] The A-3 has absolutely no glide ratio, and we'll drop like a brick onto the coastline or the tarmac."

To not further "fuel the fire", Lt. Jones didn't go into details regarding "unusable fuel", which is to say that much of the fuel left in our EA-3B's tanks would not be available for the two engines. Imagine sipping through a straw from a covered container: there's always some liquid left at the bottom unless you jiggle and turn the container while sipping harder (my grandkids somehow always get the last drop from their *"Juicy-juice"* cartons) – this can't be done with aircraft fuel systems.

TO DITCH, or NOT TO DITCH. That, *Amigo (and Amiga),* is the question.

There was, of course, nothing democratic about established military procedures. NATOPS, which is based on distilled wisdom from expert input, in-flight experiences, and engineering sheets containing factual information, dictated what had to be done by the aircrew under emergency conditions. NATOPS clearly pointed to the decision being squarely in the Pilot's purview. To come up with a rational decision, however, Lt. Jones wanted to ensure that he had factored in all pertinent information.

---

[81] Two noteworthy dual-engine aircraft flame-outs due to fuel starvation: On August 24, 2001, Air Transat Flight 236, operated by an Airbus A330, experienced a flame out on both of its engines due to fuel starvation. The plane continued to glide until it landed safely in Azores. All 306 passengers and crew on board the plane were unharmed.
On 29 November 2013, a Police Scotland Eurocopter EC135-T2+ experienced a double engine flameout and crashed into a Glasgow pub, the Clutha Vaults. Three persons in the aircraft and seven on the ground were killed; an additional 32 were injured. Both engines flamed out about 32 seconds apart due to fuel starvation.
A noteworthy VQ-1 aircraft mishap due to fuel starvation: On 16 March 1970 a Super Constellation crashed while landing at Da Nang, with five crewmen was on an over water navigational training flight from Guam to the Philippines. Unable to locate land, the crew was forced to bailout at the fuel exhaustion point. The entire crew was picked up by a helicopter from the Japanese destroyer Haruna.

"Nav – I'm going to wiggle the wings to see if that changes our fuel gauge readings."[82]

Lt. Jones turned the yoke to port, then to starboard. He repeated the action several times.

> "Not much change, Skipper. Looks like we have more fuel in Number Two tank."

Our Skywarrior had plenty of thrust, and truth be told, lacking today's ubiquitous microprocessors, there was not much in the way of an in-flight fuel conservation system. Powered by 2 x Pratt & Whitney J57-P-10 turbojets, rated at 12,400 lbf (55.16 kN) of static thrust each, the Pilot's best back-up fuel conservation source was the carbon-unit located in the brain bucket to his right. And for better or for worse, Lt. Jones wasn't putting much trust in that source right now.

"Nav – let's keep both tanks at the same level. Might increase our chances of getting a good fuel flow when we hit the low fuel warning indicator."

"Roger – fuel transfer switch "ON".

Blazer had initiated fuel transfer from the aft tank (Nr. 2) to the forward tank (Nr. 1). In the aggregate they could both hold about 3,900 gallons, enough to give our Skywarrior legs for a 2.100 mile transit, about twice the distance between Rota and Lajes. The fact that we had left

---

[82] To read military and civil airliner fuel remainders, aircraft have a few low voltage capacitors where the fuel can go between them. At a different fuel level you will have different capacitances (i.e. capacitance proportional to the height of fuel) and therefore you know where your level is at. For the aircraft pitch and roll attitudes the fuel computer works out a plain of where the fuel is (slightly different on different manufacturers). Additionally to that you have a density and permittivity readers in the tank to then calculate the fuel quantity (either kgs. or lbs.). You also have level sensors. These have two states they know, either 'wet' or 'dry'. So they are placed at known levels in the tank (high level, low level etc.) so these will trigger certain valves and warnings.

Rota with enough fuel for a 1,500 mile flight was no consolation. We had already burned through our "cushion", and were still one hundred miles out, now flirting with the low fuel warning.

"Chuck – the LOW FUEL WARNING LIGHT IS **ON**!" Nav now sounded robotic. He shuffled through his copy of NATOPS Emergency Procedures, NOMEX-glove-covered fingers flicking with cyborg precision through its pages. The NATOPS blue covers in Nav's hands brought to mind that quip I had heard from one of the para-riggers before take-off, that NATOPS stands for "Not Applicable To Our Present Situation."

"Crew – **PREPARE TO DITCH**." Lt. Jones was just as atonal. We were all in a picture show with an unknown director located somewhere outside of the aircraft's skin.

"Lajes Control, this is Ranger-17. We are prepared to ditch at sea. I repeat, prepared to DITCH at SEA. Request RANGE and BEARING to Lajes Runway One Five Zero."

"Nav – work out our coordinates, and get an SOS ginned up to send to home plate over HF (*High Frequency – which offered much greater propagation to keep in touch with VQ-2 HQ*)."

"Ranger-17, this is Lajes Control. Your location: Range 96 (*nautical miles*), Bearing 031. Your Squadron Oh-Eye-See (*OiC, Officer in Charge*) Lieutenant Commander Steve Donnelly is in the Tower. He will talk you in for your approach."

"Roger, Lajes Control. Break-Break, Stinger – this is Chuck. Glad that you're on the horn, Bubba! **My Fun Meter is Pegged, and I'm about to send this bird back to the taxpayers**. Low fuel warning for 96. Any advice?"

"Chuck, Stinger here. No sense in ditching. Too choppy. **You gotta reach for the tarmac, buddy boy!**"

"Stinger – I'm going low, in case of a FLAME-OUT. Descending to Angels TWO."

Sure enough, as we descended it was raining, and the visibility got worse after we punched below 10,000 feet.

"Skipper – with this weather we'll be lucky to get 20-mile visibility... no way to dead-reckon into Lajes. We need position updates every three mikes."

"Stinger – descending, in the Goo (*aviation slang – poor weather, clouds*), give position updates every three mikes."

We back-enders had NO idea what we were in for... Ditch at Sea; FLAME-OUT "on the rocks"; crash-land on runway one-five-zero?

"Chuck – you're at 45 miles, bearing 030. Winds east-south-east at 22 knots. You got these ZOOMIEs' (*slang for Air Force personnel*) Crash Crews all fired up. They even put some frosting[83] on the runway for you, although it won't hold because of the rain."

"Low fuel warning light is burning STEADY, Stinger. The "drink" (*ocean*) looks bad – whitecaps be damned!" Chuck was confiding in Stinger, and had completely departed from military communications procedures.

"Stinger – I don't want to go "dirty" (*aviation slang for landing gear and flaps down*) until I see the whites of your eyes – we can't afford to waste any fuel."

---

[83] The U.S. FAA recommended foam paths for emergency landings beginning in 1966, but withdrew that recommendation in 1987, although it did not bar its use. In 2002, a circular recommended against using pre-foaming except in certain circumstances. In particular, the FAA was concerned that pre-foaming would deplete firefighting foam supplies in the event they were needed to respond to a fire. Also, foam on the runway may decrease the effectiveness of the landing airplane's brakes, possibly leading to it overshooting the runway. Foam is still used in aviation firefighting, usually in conjunction with Purple-K dry chemical.

"Chuck – if you don't check landing gear at 10 out, you run the risk of turning the runway into a smoking hole (*aviation slang for airplane crash site*). Your call. You're the boss when it comes to calling your three down and locked."

"Thanks for the vote of confidence, Stinger. Remind Tower that it's my call... I don't want a wave-off... Not that I would listen!"

"Chuck – you're 40 out, bearing 030, same winds. Tower reminds me that tomorrow is Monday, so it's the Chaplain's Day Off. **Give us no surprises, OK?**"

Good leadership calls for constant communication with the troops, and Lt. Jones was red-lining when it came to LEADERSHIP.

"Crew – the sea is too rough for DITCHING. Might cartwheel, or break apart. We're going to recover in Lajes. There's a welcome party waiting for us on the Runway."

Ok, so "on the rocks" or crash-land on one-five-zero. We passengers remained strapped-in, visors down, oxygen on. Not a peep from us kids as long as Dad kept us airborne. If he were to head for the rocks, however, the sounds of gnashing teeth and "whaling" would surely reach the heavens!

"Stinger – I've got visual at 20 out. What a SIGHT! Lajes looks like a Christmas tree. Thanks for dressing up for the occasion."

All of the runway and taxiway lights had been switched on for the Emergency Landing. Just so happens that Lajes airfield is military on one side, and civilian on the other. Both crash crews were out, all of the trucks fully lit up, flashing orange and red.

"Coming to One-Five-Zero for FINAL. My TACAN has your VORTAC at 12 miles out."

"Ranger-17, this is Lajes Tower. You're 10 miles from DME, cleared for APV (*Approach with Vertical Guidance*). Godspeed."

Rain beat down on the cockpit. Blazer turned the windshield wipers on. (*Yes, Skywarriors have windshield wipers – about the same size as your HUMMER's).*

"On approach – FIVE out!"

Things were looking good now. We had to hold altitude in case of FLAME-OUT.

"On approach – TWO out, Descending to enter the glide path."

"On glide path. Will come in hard. THREE DOWN AND LOCKED!"

BAMMM!!!.... **Bam**m! The two main wheels slammed down, compressing the struts to the max... followed by a nose wheel pounding that pushed the Pilot, Nav, and PC up off their seats, rattling these front-enders' helmets against cockpit glass.

Lt. Jones had slammed the plane down hard. Struts were probably ruptured, even bent. A small price to pay for saving his crew.

For all of us this was bar-none the BEST Whale dance ever. The drogue chute deployed, and "FWOOSH" the main chute opened fully as we sped past the many assembled emergency response vehicles on both sides of the runway. Their flashing lights followed us down the runway from a safe distance, and after we reached a full stop, the fire trucks encircled the Whale with high-pressure extinguisher nozzles aimed our way.

We all celebrated with a loud and proudly tear-wrenching shout-out through the mikes in our O2 masks – a cheer that came straight from the GUT. Rooskie led a hooray chant for Skipper:

"**Hymn, Hymn, F__K Hymn!**" In the least bit melodious, to be sure, but it had a good beat, and was very apropos.

For me this ended to the longest Pucker Factor ever experienced.

"Ranger-17, Welcome to Lajes Air Force Base. If you can taxi off the runway please proceed to the AIMD hangar (*Aircraft Intermediate Maintenance Department, host for the VQ-2 maintenance detachment*)."

And expeditiously taxi off the runway we did, glad to be alive, ready to devour the north end of a south-bound whale. Tomorrow would be a down-day as AIMD replaced struts and got Ranger-17 ready for Bear hunting.

Enter Aries – Ram Meets Bear in Baltic

It was commonplace for Airborne Cryptologic Technicians to "progress" from EA-3B Skywarrior aviation into EP-3E Aries aviation (by the way, "Aries" stands for **A**irborne **R**econnaissance **I**ntegrated **E**lectronic **S**ystem – *cool*!). For me this progression occurred within the same VQ-2 Squadron based in Rota, Spain (and, for the record, this experience I'm describing was similar for many of the Pacific Fleet Airborne Cryptologic Technicians who flew with VQ-1 based in Agana, Guam). I think that it's worth pointing out that these two aircraft, while belonging to the same squadron, and intended to conduct similar missions, had significantly different capabilities, different operational profiles, and different organizational cultures surrounding them. The original task of the EP-3 aircraft was to eavesdrop on Warsaw Pact Navies, fingerprinting their radar and communication suites.[84]

---

[84] See http://www.navair.navy.mil/index.cfm?fuseaction=home.displayPlatform&key=5B7DC7D0-C4DB-45E3-92EA-662605B57409 and http://www.spyflight.co.uk/ep3.htm. Both retrieved on Dec. 22, 2016.

When you look at the difference in crew size alone (Skywarrior 7 vs. Aries 24 to 28[85]), you realize that not only are the Team dynamics different, but the SIGINT output is itself different in terms of sheer volume of product, and the level of detail that obtains from more sophisticated in-flight analysis.

A tactical SIGINT report produced on the Skywarrior was intended to instantly alert the Battle Group Commander about changes taking place in the surrounding tactical environment. A tactical or national SIGINT report produced on the Aries EP-3E (*sometimes affectionately referred to as Sky Pig, in the same vein that the Skywarrior was affectionately called the Whale*), however, would support not only the Battle Group Commander when the bird was operating in a Fleet Support mode, but also the Commander Sixth Fleet staff, the Commander in Chief of Naval Forces Europe, and National consumers as well.

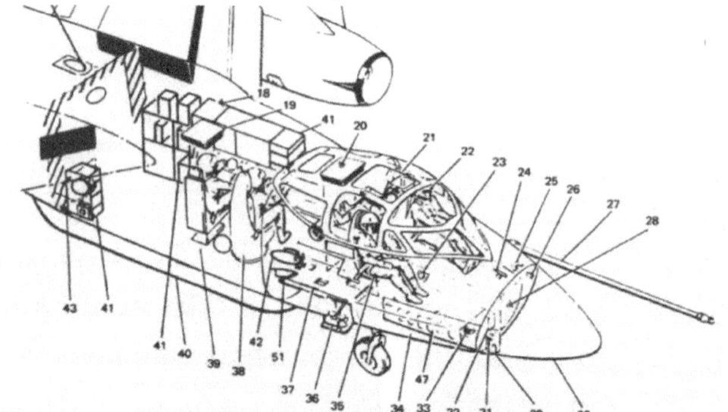

Aries-produced reports would go out not only over tactical voice communications channels, they would also go out in textual record traffic over national communications circuits, and thus serve to tip-off other national SIGINT collection resources. These national SIGINT communications circuits were operated by yet another type of Airborne Cryptologic Technician – the CT-Communicator, whose rating designator was "CTO", commonly referred to as the "O-Brancher" *(just as CTI*

[85] This de-classified CIA document provides a good description of EP-3E Aries-II ELINT aircrew organization and functions – it is silent on COMINT aircrew organization and functions: https://archive.org/stream/CIADocuments/CIA-531_djvu.txt. Retrieved on Dec. 22, 2016. (see Appendix)

Interpreters were "I-Branchers"; CTR Morse and Signal Collectors were "R-Branchers"; CTM Maintenance Techs "M-Branchers"; CTA Admin-Techs "A-Branchers; CTT Special Signals Techs "T-Branchers", or simply "T-birds").

In addition, the EP-3E often operated in a National SIGINT

Appendix F (U) Schematic of EP-3E with Position Identifications

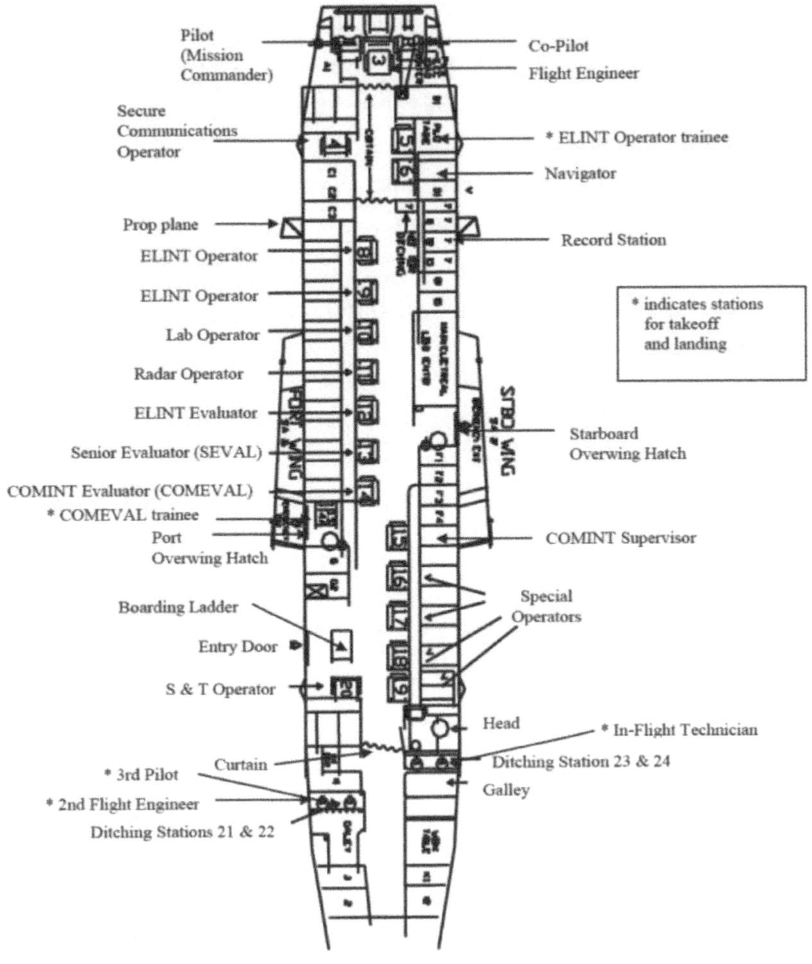

Collection mode, meaning that it operated over waters from where it could gain SIGINT that was of interest to National decision-makers. These missions were flown under the purview of the Peacetime Aerial

Reconnaissance Program (PARPRO)[86] following a complex set of rules designed to minimize risk, while maximizing oversight by and accountability to the Joint Chiefs of Staff and by/to the National Intelligence Community. Most of the time our Aries-II missions were owned by the Intel Community and paid with National Intelligence funds, therefore flown under PARPRO sponsorship for the benefit of the intel Community at large.

It's also evident that the Carrier Battle Group aviation culture was neither employed, nor necessarily understood by the Aries clan. Gone were the BOLTERS, the wave-offs, and the $2 Cats and Traps. HELLO to, and welcome became the cushy international deployments with hotel accommodations in foreign lands, and the getting used to a *"Semper Per-Diem"* (Latin for *"daily $upplement forever"*, since we received a daily pay allowance when room and board had to be procured from the local economy, rather than at a military facility) mentality.

These new creature-feature "bennies" (*job perks*) notwithstanding, Aries aviation still presented dangers, since the lone wolf aerial reconnaissance profiles we flew alongside coasts that harbored potentially hostile militaries frequently triggered responses from their Air-Defense Aviation.

---

86 http://airforcehistoryindex.org/data/001/051/427.xml. Retrieved on Dec. 22, 2016. Recently declassified USAF document lists the some of the Air Force PARPRO component operations: PEACETIME AERIAL RECONNAISSANCE PROGRAM TASKED GIANT CLIPPER, GIANT SCALE, GIANT REACH, OLIVE TREE, OLIVE HARVEST, SENIOR LOOK, OLYMPIC GAME, OLYMPIC TORCH, OLYMPIC VICTOR and OLYMPIC FIRE MISSIONS.

It was normal for Soviet Air Defense fighter aircraft to "react" to our PARPRO flights, and their Baltic Air Defense bases, the most heavily armed Soviet region being Kaliningrad[87], would send SU-15 "Flagons" out to use us as targets-of-opportunity, and to escort us outside of their Air Defense Zones. The Flagons always flew in pairs, and would consistently take turns in approaching our aircraft from below, and from the rear, in order to optimize their Anti-Air missile kill profiles, since they carried AA-8 "Aphids" with heat-seeking homers[88]. Today these same kinds of encounters are commonplace off the coast of Kaliningrad, only now the aircraft are more sleek and capable – case in point a recent account from the Pentagon about our US Air Force RC-135 Rivet Joint SIGINT

---

[87] http://www.numbers-stations.com/articles/russian-military-in-kaliningrad/ Retrieved on Dec. 22, 2016. Kaliningrad — former East Prussia of Germany is in Russian possession since 1945 when according to Potsdam conference Eastern Prussia was divided between Soviet Union and Poland. The regional capital Königsberg was renamed Kaliningrad. As the westernmost territory of Soviet Union it gained its military importance primary for Soviet Baltic navy as a spearhead into the Atlantic. After the collapse of the Soviet Union Kaliningrad became part of Russian Federation was separated from rest of Russia as Lithuania regained its independence. Today Kaliningrad region is sandwiched between Lithuania and Poland. This increased the military importance of the region as the Baltic Navy struggled to keep its position in the Baltic Sea and compensate the loss of Liepāja War port that was regained by Latvia. Since 2014 following rise of tensions between Russia and NATO countries, Kaliningrad region gained a prime importance in the Baltic States security question. Western and Russian military experts view Kaliningrad region as important bridgehead for Russian military and navy to gain control over Baltic States and Baltic Sea.

[88] For a comprehensive description of SU-15 Flagon capabilities, go to: http://all-aero.com/index.php/contactus/10740-sukhoi-t-58--t-59--t-60--su-15-flagon. Retrieved on Dec. 22, 2016.

Reconnaissance aircraft being intercepted by Russian Air Defense Forces SU-27 Flankers, causing an "Incident in the Sky".[89]

For your edification, consider this "routine" Aries PARPRO mission off the Eastern shores of the Baltic Sea, a mission that took place in the fall of 1979. Luckily we were flying one of the newly upgraded Aries-II birds, one equipped with the "DEEPWELL" ALR-60 computerized COMINT collection system. Over the course of the year I had qualified as one of three VQ-2 certified DEEPWELL ALR-60 Instructors, and knew that when we used it effectively, we could squeeze the juice out of that DEEPWELL collection system to come up with some real eye-popping Intel scoops. During this particular mission we would bag some pearls the likes of which had never been seen in Navy PARPRO missions when targeting Soviet Baltic Fleet operations.

The importance of Soviet Baltic Fleet operations derived from the fact that, from an historical perspective, the Baltic Fleet was considered pre-eminent because of the role that this Navy's presence in St. Petersburg had played in the 1917 October Bolshevik Revolution (*which by the way was dated to 25 October 1917 by the Julian or Old Style calendar, and corresponds to 7 November 1917 in the Gregorian or New Style calendar – i.e. what the rest of the world uses*), and from a geo-

---

[89] http://thesentinel.ca/ru-su-27-intercepts-us-rc-135u-recon-aircraft/ Retrieved on Dec. 22, 2016. On the morning of April 7 2015, a US RC-135U, flying a routine route in international airspace, was intercepted by a Russian Su-27 Flanker in an unsafe and unprofessional manner," the Washington Free Beacon (WFB) cited Pentagon spokeswoman Eileen M. Lainez as saying. "The United States is raising this incident with Russia in the appropriate diplomatic and official channels," she added. The US RC-135U was most likely monitoring Russia's military activity in Kaliningrad. Since the recent deployment of Iskander mobile short-range ballistic missile systems in the Kaliningrad Oblast, NATO has been closely monitoring the situation. NATO has built up a sizeable force along Russia's western border. Adding to that, NATO's Response Force has seen its numbers bolstered to 30,000 troops and have been conducting many joint multinational training in the Baltic States. Russia warned NATO that recent intensification of military exercises in the Baltic States was "unprecedented and dangerous."

political perspective in that it had an important role to play in the USSR's anti-NATO strategic plans, which assigned important amphibious assault operations against Denmark and Germany to this "Twice-ordained Order of the Red Banner" Baltic Fleet (*Krasno-znamyonnij Baltiskij Flot*). In addition, the Baltic Sea offered convenient staging waters to put new maritime weapons systems to test, with sea trials usually being held in the summer and fall, prior to the winter cold setting in. Naval exercises abounded, and were always signaled by "NOTAMs" (*Notices to All Mariners*) that the USSR issued to the ICAO (*International Civil Aviation Organization*) in order to declare area closures in the Baltic Sea. These NOTAMs were warnings designed to keep foreign military and commercial ships and aircraft from intruding on sensitive Soviet military operations. We did the same throughout many parts of the world.

RAF Sqn. 51 Nimrod

Much to my good fortune, VQ-2 Aries aircraft were used as one of the principal U.S. SIGINT platforms when it came to gathering information on the Order-of-Battle of Soviet military forces in the Baltic area-of-operations. Other NATO military powers also kept close tabs on Soviet military activities in the Baltic, to include Sweden, Norway, Denmark, Germany, and the United Kingdom, whose Nimrod SIGINT birds frequently flew the same kinds of SIGINT collection tracks that we Aries aviators enjoyed. I had committed to becoming a credible Baltic Fleet analyst, and at the time had more than four Baltic-deployments under my belt. My name was permanently in the hat for any and all upcoming Baltic missions. This time I had been wrenched away from Victoria, who was pregnant with Paul, and was due to give birth within the month. As one of only two qualified Baltic COMINT Crew Chiefs I had been pressed for duty because my counterpart, CTI1-Russian linguist Mike Caldwell was back in the USA on a family emergency. Victoria and I were banking on my return in time for Paul's birth at the Naval Hospital in Rota, Spain. Just in case, to keep Victoria company, her mother had arrived from Madrid and was staying with her.

"Mak – I want you to do the Intel Brief for tomorrow's mission. Our pre-flight van will depart the BOQ (*Bachelor Officer Quarters*) at 5:30." Gunner Hank Stark, a Marine Warrant Officer, was gruff, short on words, but long on trust and leadership. He had somehow found a way to qualify as an EP-3E Communications Evaluator (*COMEVAL*), although his background was in nuclear weapons security. Given his seniority, this might have been his choice for much deserved swan-song duty before retiring from a 30-year career.

"Got it, Gunner. I'll head over to the Intel Center before turning in to get a leg up on FOSIF forecasts." I was junior to the CTI1 Polish linguist, PO1 Walt Smythe, but Gunner wanted the brief to focus on Soviet intel, and knew that he could get a Baltic Fleet analyst perspective from me. I had already agreed back at home-plate, when Senior Chief-CTRCS Orndorff put the COMINT team composition together, to coordinate closely with Walt on his Poland input.

The read boards described how Soviet naval forces from the Northern Fleet and the Black Sea Fleet had joined up with Baltic Fleet counterparts to form an unusual massing of an amphibious assault armada. Led by the aircraft carrier Kiev, the armada included several of the USSR's most capable amphibious landing ships, such as the Ivan Rogov[90], the largest of its kind in the Soviet Navy. These amphibious landing ships were capable of carrying two regiments of naval infantry. A regiment has 2,038 men, divided into three infantry battalions of 409 men each, a tank battalion with 31 light and 10 medium tanks, a battery of rocket launchers, and a battery of antitank weapons, as well as air defense, signal, engineer, and support companies. FOSIF reports also disclosed several NOTAMs in the Baltic, off the coasts of the USSR and Poland, one of those near the Polish port city of Gdańsk, probably intended for several amphibious landing exercises. The report quoted NATO sources (Denmark) with having obtained an accurate count of 10 Soviet Amphibious Landing Ships having entered the Baltic through the Kattegat Strait over the last three weeks.

Somewhere in the fine print, amid the factual assessments, there was speculation as to why the Soviets were holding this massive Amphibious Landing exercise in coordination with Polish and East German Warsaw Pact allies. The anonymous analyst postulated that the display of strength was intended to counter a recent visit by Pope John Paul II to

---

[90] The Ivan Rogov Landing Ship was built as part of the expansion of the Soviet Navy's amphibious warfare capabilities in the 1970s. The amphibious ship Ivan Rogov, which entered the fleet in 1978, can carry 550 naval infantrymen, tanks and other weapons, and is the first Soviet amphibious ship able to carry helicopters. The ship can launch landing craft or can drive her bow up onto the beach for a direct landing.

Warsaw, Poland. [91] Needless to say that the Pope's historic, and YES, **heroic** return to Poland in June of 1979 triggered a shit storm in the Politburo, and, our analyst contended, this "tear at the Warsaw Pact seams" had to be countered by showing the people of Poland (and of Warsaw Pact nations writ large), that Kosygin was still boss in that part of the world. What better way to do that than with an Amphibious Assault exercise on the coast of Poland near Gdańsk? This militaristic response to Poland's new lungful of religious freedom was so hastily planned that the USSR in fact violated the Helsinki Final Act of notification of military exercises.[92]

---

[91] http://catholicstraightanswers.com/pope-john-paul-iis-role-fall-soviet-union/ Retrieved on Dec. 22, 2016. On June 5, 1979, Pope John Paul II arrived in Poland to visit his homeland. As he descended the stairs of the plane in Warsaw, he kissed the ground and incited a spiritual earthquake. During his visit at the Auschwitz concentration camp, the symbol of the evil of totalitarianism, he told the thousands of people gathered from the Eastern European countries to resist the falsehoods they had been told: "You are not who they say you are, so let me remind you who you are." He preached about the dignity God had given to each of us, especially through our Savior. At Krakow, he again emphasized the need for spiritual and cultural renewal, preserving Poland's strong faith, and the transforming power of Christ's love. In response, the people chanted, "We want God, we want God, we want God in the family, we want God in the schools, we want God in books." The visit emboldened the Solidarity movement, and by 1980, the government recognized Solidarity as the first independent trade union in the communist bloc. At the time, President Reagan commented to a friend, "I have had a feeling, particularly in the pope's visit to Poland, that religion may turn out to be the Soviets' Achilles' heel." *// The author is conflating the origins of this "separatist" socio-political movement with the Soviet military exercise called "Zapad-81", by moving this event up in time by about 18 months.*

[92] http://www.britannica.com/EBchecked/topic/260615/Helsinki-Accords. Retrieved on Dec. 22, 2016. The Helsinki Accords, Helsinki Final Act, or Helsinki Declaration was the final act of the Conference on Security and Co-operation in Europe held in Finlandia Hall of Helsinki, Finland, during July and August 1, 1975. Thirty-five states, including the USA, Canada, and all European states except Albania and Andorra, signed the declaration in an attempt to improve relations between the Communist bloc and the West. The Helsinki Accords, however, were not binding as they did not have treaty status.

I got to the Bachelor Officers Quarters (BOQ) at 5:20 a.m., wearing my zoom bag (*flight suit*), my Winnebago (*USN-issue leather flight jacket*) emblazoned with a healthy collection of carrier aviation's ship and squadron patches, and a Navy ball-cap sporting PO-2 chevrons. Gunner was waiting for the bus, along with Lieutenant Commander Thoreau, today's mission commander. I snapped a brisk salute.

"Good oh-dark-thirty, PO-2 Mak. Any good scoop from FOSIF?" Gunner and LCDR T returned my salute from under their *piss-cutters* (caps worn by Officers and Senior NCO's, more lewdly referred to as *c..t-caps*).

"The Soviet Bear is spending lots of rubbles on this one, Gunner. Kosygin wants to apply some fear-glue to all of the unravelling going on in Poland." I was not prone to sarcastic quips, but at oh-dark-thirty it's too early to apply filters.

"There's a good chance that this Det. might get extended." LCDR Thoreau said. "Rumor is that interest in this activity might reach the CNO's (*Chief of Naval Operations*) boss (i.e. **Commander in Chief of the United States Armed Forces – the President of the United States of America**). Good grist for the Detachment's rumor mill, which was *constantly* at work.

We shuttled to the EUCOM Intel Center and got to work. We would later close doors on Ranger-25 at 7:10 a.m. for a 7:30 a.m. take-off (*Germany time was Zulu +2, so 5:30 a.m. Zulu*).

Echterdingen Army Air Field

The airfield we used near Stuttgart was just over 25 miles away from our EUCOM Vaihingen operations center. It was called the Echterdingen Army Air Field, and has a rather gruesome past, one that we Navy aviators were oblivious to at the time.[93] As our shuttle drove us

---

[93] During WW-II Echterdingen served as a Nazi Luftwaffe Air Field. In October 2005, workers rebuilding the entrance of the field, as part of an upgrade for new security standards, found a shallow mass grave, and exhumed human remains.

across the aircraft ramps lined with A-10 Thunderbolt-II "Warthog" tank killers, Ranger-25 became visible in the early morning haze, our trademark Sandeman looming ominously on the tail. A fuel truck was parked alongside, its hose being disconnected in preparation for take-off.

We thanked the ground-pounder for the ride, and climbed the ladder into the Aries' cavernous crew cabin. It was lined on both sides with floor-to-ceiling racks of SIGINT avionics kit, dwarfing the flight avionics gear by a nearly two-to-one ratio, perhaps more if the exterior bumps and warts for SIGINT antennae were included. One of the more remarkable features of Aries-II was its BRIGAND[94] passive ELINT collection system, integrated into the Big Look Radar antenna, and housed in an underbelly pill-shaped cowling, so sensitive and versatile that its true performance parameters remain classified.

The 34 bodies found are believed to be slave laborers who were prisoners at the nearby Echterdingen subcamp of Natzweiler-Struthof concentration camp. The prisoners at that camp were predominantly Jews, rounded up from across Europe to be worked to death under Nazism's "racial" policies. The camp provided labor details from among its weak, starved, sickly inmates to the airfield -- then, as now, a civil airfield that doubled as a military air base -- from November, 1944, to at least February, 1945. In early 1945, the airfield was seized by French troops under General Leclerc, and later handed over to Americans as the Western Allies adjusted their occupation sectors. The German police opened an investigation into the deaths of these 34 poor wretches. When evidence emerged that at least some of the victims were murdered, and two or three of them actually buried alive, prosecutors designated it a homicide investigation. Reference: http://www.aeronews.net/index.cfm?do=main.textpost&id=ca825491-694e-4a85-8da0-ec7ced6c70ad. Retrieved on Dec. 22, 2016.

[94] See http://p-3publications.com/PDF/AirborneLog-Summer93.pdf retrieved on Dec. 22, 2016 for a description of the EP-3E Aries-II SIGINT avionics suite, to include a schematic showing the BRIGAND operator's workstation. This article from Aviation Week & Space Technology is a reporter's account about his flight on a U.S. Navy EP-3E surveillance plane, like the one that was forced to land in China after colliding with a Chinese PRC fighter on April 1 2001. It describes the flight, how the aircraft is equipped and what each of the crew members does. It was originally published in May 5, 1997: http://mwaviation.tripod.com/ep-3e.htm. Retrieved on Dec. 22, 2016.

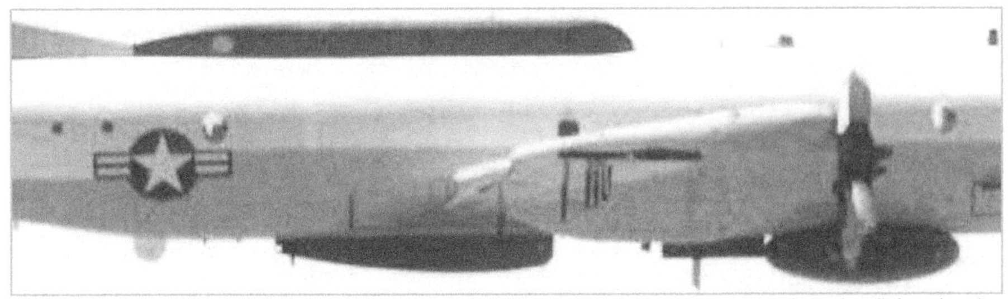

Brigand is in the pill-shaped cowling

The baseball card version of Aries-II/DEEPWELL SIGINT capabilities goes something like this: an array of direction finding, signal analysis, and recording equipment that includes the ALQ-110 “BIG LOOK” radar signals gathering system, ALD-8 radio signal direction finder, ALR-52 automatic frequency measuring receiver, ALR-60 DEEPWELL automated COMINT collection system. The aircrew included 7 officers and 17 enlisted personnel, comprising 3 pilots, a navigator and a flight engineer for flight operations from the cockpit, the remaining 19 covering ELINT and COMINT tasks using workstations in the main cabin. On important missions such as this particular Baltic series, a REWSON (**R**econnaissance, **E**lectronic **W**arfare, **S**pecial **O**perations, **N**avy.") engineer would join the party and configure the Special Signals workstation with real hush-hush gear.

We held the pre-mission brief in the spaces where the ELINT in-line and the COMINT in-line intersected, an area mid-aircraft where the wings attached to the fuselage. The Mission Commander, Mission Evaluator, COMINT Evaluator, ELINT Crew Chief and COMINT Crew Chief attended to collectively put their brains around the upcoming challenges, all based on the latest and greatest intel that had just been gleaned at the Ops Center. In addition, the O-Brancher would often receive new intel tip-offs over his comms circuits before take-off, and the updates would hastily (and gratefully) be factored into mission planning. These gatherings were not like the relatively orderly Ready Room briefings that Skywarrior crews held on aircraft carriers. Aries-II pre-missions took place while the pilots were revv’ing up the four turbo-prop engines to their maximum *ROAR*, all to ensure performance in flight, and the Flight

Engineer and Nav were running up and down the "tube" conducting systems checks, removing faceplates and floorboards to get to key switches – indicators – gauges. The best way to hear through the din was to wear a pair of headsets, and sometimes even then, earplugs (*"frog-dicks"*) had to be applied.

The Mission Commander always went first with track information, take-off time and ETA, and a recap of higher-echelon tasking. The Mission Evaluator would hi-light expected radar hits from selected high-priority target platforms. The COMINT evaluator would describe likely fighter reactions from pertinent Air Defense Zone fighter bases, and describe the rationale behind the CTI Linguist mix (in this case Russian – 3, Polish – 1, and German – 1), and describe the CTR Manual Morse operator's function in monitoring Soviet, Polish, and East German air defense tracking networks. When there was a complex Soviet naval exercise in the offing, the COMINT Crew Chief would brief whatever distilled pertinent wisdom was contained in the most recent intel reports.

"Now get to work! We'll be on track within an hour. If you have an equipment gripe let me know and I'll prioritize it with the in-flight tech." LCDR Thoreau had about three thousand hours on EP-3Es, and knew the Baltic, the Med, and North Atlantic like few others in the Squadron. Today we were positioned to hit a home run, given the onboard talent, and the unfolding Soviet amphibious landing exercise we were about to dissect.

For our CTI1 German linguist the "track" started much sooner, since he could hear well into East German territory as soon as we gained cruising altitude. CTI1 Karl Rauschenberger was, as his surname indicates, of German descent, and had grown up in Ramstein, one of the many U.S. Strategic Air Command (SAC) bases in Germany, where his dad was employed with the 86th Air Base Group. Karl was a "Deutschland über alles" kind of a guy, a personality trait that would always trip CTI1 Walt Smythe's wire when we dined at the local Gaststätte Hondler, about half a mile from Patch Barracks.

"Look, Walt – it's not only that we Germans have the best *braü*, we also have the best ***fraü*** *(-en)* which gives us unabashed bragging rights." Amazing what a liter of *Schwaben Bräu* would do to loosen up the inhibitions, as Karl winked at the svelte blond waitress named Karina, porting his next half-liter round. And Karl could sure talk it up with the locals – you name it, soccer (*Fußball Sie dummkopf!*), politics, history, anything and everything . We were lucky when he stuck with us for our evenings about town, since he was usually flying solo, philandering *mit den Frauen* on easy-come, easy-go per-diem greenback.

"Karl – I concede that the food and eye candy are hard to beat, but just scratch the surface, and you'll find that the same Germans that gave rise to Hitler can easily do it again. It's in their blood – what racists like de Gobineu and Houston Chamberlain (*Richard Wagner's son-in-law*) put down on paper about the "Master Aryan race" these people bought into lock, stock, and barrel!"

"Ahh – bullshit, Walt. Germany was great before any of that master-race crap ever kicked in, and it's great today! It's all about Teutonic mythology – it brings out the sparkle and promise of humanity against chaos!" Almost in the same breath, had he not …*GLUB*… chugged heavily from his tall stein of *Schwaben Bräu*, he challenged Walt: "I bet you next round that I can get this vixen amazon's phone number before the bill comes."

"You're on, Karl." Walt had sized Karina up, and knew that the odds were in his favor. Besides, Walt was always keen for a free brewski.

Karl winked at Karina, and signaled her over:

"Bitte schatz, welche Zahl sollte ich anrufen, um eine Reservierung für nächste Woche zu machen?"

Karina smiled, took out a piece of paper, jotted down a phone number, and said in perfect Queen's English:

"Best to call in the afternoon before all the times are taken." She winked back at Karl.

Touché! Definitely Walt's round coming up.

What Walt didn't know was how to interpret the *spiel* that Karl had sprung on Karina:

"Please sweetheart, what number should I call to make a reservation for next week?" – Language: the key to communication. If you don't know the language, you're definitely at a loss.

Now, at Poz-13, Karl was dutifully working his language, carefully sorting through the comms signals that were bursting like popcorn from out of his receivers. Much of the comms chatter came from the *Kommando der Volksmarine* (People's Navy HQ Command) in Rostock-Gehlsdorf. The HQ was in charge of a small, but capable Navy whose main mission was to counter NATO forces, and its secondary mission was the prevention of "*republikflucht*" (people leaving the GDR without official permission). The East German People's Navy was comprised of:

1st Flotilla in Peenemünde,

4th Flotilla in Rostock-Warnemünde,

6th Flotilla at Bug on Rügen Island,

6th Border Brigade (Coast) in Rostock.

Karl filled-out a report template, and passed it down the line for me to take up with Gunner Stark. Karl kept gisting away. The report read:

"05:58Z – East German KV tasked an undetermined number of Osa, Tarantul, and Shershen fast attack craft of the First and Fourth Flotillas to participate in coastal defense exercises scheduled from 12:00 Moscow time (Zulu +4) to 20:00 Moscow time in an area defined by coordinates in Rostock-Gehlsdorf NOTAM of 21-09-79."

"Gunner – the East Germans are already spooled up for coastal defense exercises today, starting as early as 08:00 Zulu, two hours from now. I recommend release." I handed the report over to Gunner, who then met with the Mission Evaluator to add any ELINT that pertained. There was nothing, and Gunner released the report for immediate transmission via secure SATCOM.

"We're ON-TRACK" LCDR Thoreau announced over the All-Crew ICS.

The track consisted of a dog-leg that extended from an overwater point Northwest of Peenemünde, along the Polish coastline to a point about 100 NM North of Kaliningrad, where it veered further North along the Lithuanian and Latvian coastline, reaching a point East of Ventspils, where we reversed course to backtrack. The bird had enough fuel to travel the full length of the track four times over (what in swimming terms would be two laps), but in reality it was up to the Mission Commander, in consultation with the Mission Evaluator whether to "orbit" around areas of primary interest. Today that area of primary interest was centered around the Gdańsk – Kaliningrad segment of the track. There was no reason to go North of Kaliningrad unless we found little or no activity there on our first pass.

"Walt – when was the last time that Polish fighters reacted to one of our NATO *Rekky* (*short for reconnaissance*) birds?" Gunner needed to know.

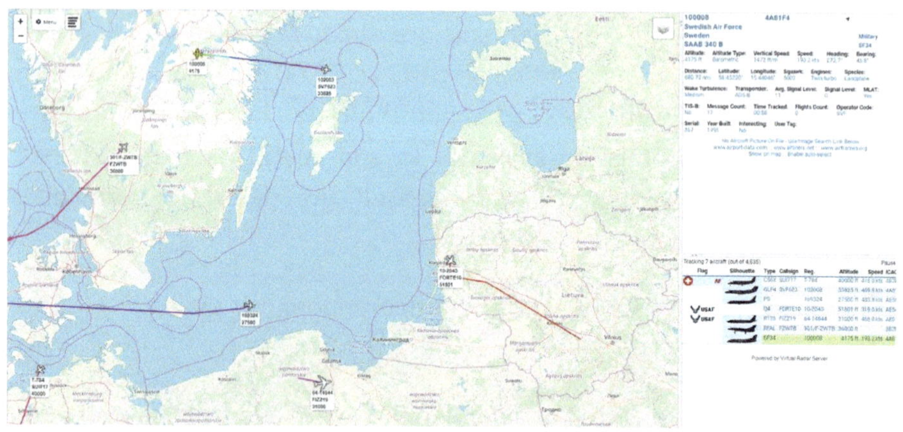

"Polish fighters very seldom react overwater – they primarily train over land against Soviet fighters. There's a **Soviet** Fighter Regiment at Kolobrzeg with some SU-15 Flagons[95] who use NATO *Rekky* birds as targets of opportunity. They could easily do a number on us. You probably saw in the news last year that It was a Flagon that bagged the Korean airlines flight 902 (*KAL-902*[96]) for violating Soviet airspace. My

95 The Soviets were in control of Eastern Europe's airspace. In strategically located (relative to the main potential theater of operations along the West German frontier) Poland, the USSR based combat aircraft at Legnica, Gniezno, Pucza Bolimowska, Gdansk (naval air Backfires), Zagan, Brzeg, Opole, Szczecin, **Kolobrzeg**, Szczecinek and Koszalin. The bases at Zagan and Szprotawa hosted Su-24 Fencer aircraft. See Robert E. Harvaky's "Bases During the Cold War" to understand the NATO vs. Warsaw Pact chess board: http://img.rtp.pt/mcm/pdf/840/84048a68b8df843f71c53c161fef5f071.pdf. Retrieved on Dec. 22, 2016.

96 Pearson, David Eric (1987). KAL 007: The Cover-up. N.Y.: Summit Books. ISBN 0-671-55716-5. On 20 April 1978, Soviet air defense shot down Korean Air Lines Flight 902 (KAL 902) near Murmansk, Soviet Union, after the civilian aircraft violated Soviet airspace and failed to respond to Soviet ground control and interceptors. Soviet Air Defense Forces initially identified it as part of the US air reconnaissance force, which carried out thousands of flights along Soviet borders annually at the time. Captain Alexander Bosov, pilot of the Sukhoi Su-15 that brought down Flight 902, saw Asian logogram characters on the tail of the Korean aircraft, and reported this to the ground control. Despite this, Vladimir Tsarkov, commander of the 21st Soviet Air Defense Corps, ordered Bosov to take down the plane, as the plane failed to respond to repeated orders to land, and

recommendation to Nav is that we stay on track, and that we don't stray any closer to Warsaw Pact airspace."

Walt was talkative, and mindful of the perils that drifting East entailed. The rules of PARPRO required that we maintain our distance well outside of territorial waters, and moreover, outside of the reach of the many Air Defense Missile batteries, who for the most part had a kill radius of about 30 miles from the coast. If we came to within what was still considered to be very prudent distances from the territorial waters demarcation line, we were required to IMMEDIATELY report our risky position, and head back out toward open ocean while taking pre-emptive evasive measures.

"Mak – how about you concentrate on Fighter Reaction for the duration of the mission. All of these bastards are Soviets, regardless of where they take-off and land."

"Roger that, Gunner. I got it covered." Things were pretty much routine as we traversed North of the Polish coastline. Kolobrzeg was quiet today.

Not that air defense fighters were my kettle of fish – too easy. Very predictable comms patterns, and no complex terminology, but it

---

was approaching the Soviet border with Finland. The Su-15 opened fire, forcing the plane to descend, and killing two of the 109 passengers and crew members aboard Flight 902. The plane made an emergency landing on the frozen Korpiyarvi Lake (not to be confused with the Korpijärvi Lake) near the Finnish border. Also see: Gollin, James; Allardyce, Robert (1994). Desired Track. The Tragic Flight of KAL Flight 007. American Vision Publishing. ISBN 1-883868-01-7, describes how a Soviet PVO's Flagon downed a Korean airliner on September 1, 1983. The South Korean airliner serving the flight was shot down by a Soviet Su-15 interceptor, near Moneron Island west of Sakhalin in the Sea of Japan. The interceptor's pilot was Major Gennadi Osipovich. **All 269 passengers and crew aboard were killed**, including Larry McDonald, a Representative from Georgia in the United States House of Representatives. The aircraft was en-route from Anchorage to Seoul when it flew through Soviet prohibited airspace around the time of a U.S. aerial reconnaissance mission.

was customary for the more seasoned Russian linguist to saddle up against what we all knew posed our foremost existential threat: Soviet Aviation PVO (*Protivo-Vozdushnoy Oborony*, or Anti-Air Defense) fighter aircraft. To crusty *Rekky* aviators (*spelled RECCE*), the mere sound of the acronym *"PVO"* would conjure images of pythons and vipers lashing out with deadly venom to take-down less nimble passers-by. A navigational error could and would easily trigger a KAL-902 experience for us.

I preferred the challenge of large-scale open ocean operations such as this MASSIVE amphibious landing exercise that was in the offing near Gdańsk. Our two other CTI Russian linguists were going to have their hands full managing the morass of ship-to-shore and ship-to-ship comms originating from about 20 surface combatants, their respective headquarters, their respective Officers in Tactical Command, and at least a dozen coastal defense tracking posts (*what we called NIS-Posts, the NIS standing for Nablyudeniye I Svyaziy, which roughly translates to Surveillance and Communications*), not to mention the range closure security ships, who chattered and barked like pissed-off yard-dogs. With DEEPWELL acting as a COMINT system force multiplier, Ranger-25 had a great "signals collection" advantage in that its central processor would instruct receivers and recorders to automatically log and record ALL the activity that came up on the various ship-to-ship, ship-to-shore, NIS, and other pertinent communications networks. To facilitate a timely review of these recordings, DEEPWELL operator interfaces provided all of the recording log information, and linguists could bring-up and process these recordings in-flight. That said, it was the post-mission reconstruction of events through a careful ground-processing operation that would yield true "A-validity" intel for our warfighters. That careful analysis would start back at Rota, where all the recordings were carefully transcribed in playback mode on a sister (but dumbed-down) ALR-60 DEEPWELL system, and continue throughout specialized intel centers within the global Intel Community network.

In the "PVO" saddle it was my responsibility to work closely with CTR2 Jack Tindall, who had been plotting Soviet Air Defense tracks since before we went "feet wet" at 06:11 a.m. Zulu by leaving the German landmass behind to thunder-on into the Baltic Sea. Jack had been following exactly how the Soviets tracked our aircraft as we proceeded Northeasterly up the coast of Poland. The Morse code data that he plotted on his grid-map of the Baltic Sea focused almost exclusively on where it was that the Soviet Air Defense Command saw **us** on the map. When the pot was percolating, so to speak, the Soviets' tracks were consistently about 2 to 3 minutes old, but good enough for the various Ground Control Intercept (*GCI*) stations along the coastline to vector fighters directly at us.

"Jack – how well does the Soviet data jibe with Nav's data?"

"Pretty close, Mak – looks like they're not too concerned about us yet, since they're only updating our track about every 15 mikes."

"So how far off were they from Nav's data... five miles, ten...?"

"About 10. Chances are that the closer we get to one of their GCI centers their long-range ground radar triangulations will become more accurate."

So far, so good... we would know when the Soviet Air Defense HQ was more concerned about us if and when they started reporting our track every THREE minutes.

"Hey Mak" – a tap on my left shoulder – I thought it was Gunner. No one to my left. I looked right...

"No situational awareness –eh? Whaddaya say about chipping in the five bucks you owe for steak and spuds, ya numb-nuts Spook?" Sure enough, it was AW2 Larry Bunker. He always "volunteered" to take on the onerous task of ensuring that the entire crew got at least one decent meal during our ten-to-eleven hour flight. His only qualification for meal preparation was that he had worked as a short-order cook at a Sizzler's restaurant before he enlisted in the Navy in Duluth Minnesota. His nick-name "Slap-shot" told volumes about his days at high school hockey in

Minnesota. When he was busy in the Galley, however, he was "Crock-pot".

"Crock – what a crock-o'-shit to question my SA. Here I am, spinnin' the dials to keep a tab on fighter-interceptors coming after us, and you want me to pay for my meal? Should be on the house!"

"Pay up, you frikkin' deadbeat, or face KP (*kitchen patrol*)!... Hey Gunner – we got a mutineer at Poz-11." I grudgingly forked over the $5, knowing that if I par-boiled my flight boot I might get about the same results as Crock. All jest aside, Crock could really put together a square meal, and no one could figure out how he did it... some steaks, some garlic salt, some spuds, some frozen veggies, maybe a lot of butter. Chow was often ready right about the time when the entire ELINT In-line, and the entire COMINT In-line was assholes and elbows, fully engaged in red-hot surveillance. It was then when the wafting aroma of charred steak would permeate the cabin. The pilots always ate first, then whoever happened to have a light load would make his way back to the Galley, and so on, until Slap-shot got to feast on what was left. Sometimes Slap-shot would commandeer a trainee who was sitting in a Siesta Seat (*there were 4 seats aft of the cabin door for trainees*) memorizing NATOPS or Soviet weapons systems, to deliver the meals, a la flight-attendant. As for drinks, whatever you brought in your helmet bag stayed on the floor by your feet, and if it didn't freeze, you were good to go. Water and NAVY coffee (*the kind you could stand a spoon in*) were ALWAYS available in the Galley.

"Gunner – DEEPWELL shows that there's some activity on an Air Defense PVO airfield frequency. I'll check the logs to see who might be coming out."

I searched the DEEPWELL recording logs, and sure enough, there was a pair of Mig-23 Floggers active on their local airfield frequency at Ventspils. When I gisted the recording, I found that two Floggers (presumably F-variant PVO fighter-interceptors, which is what my Order-of-Battle tech kit indicated at Ventspils - although the call-signs did not equate) had been doing touch-and-go's at low altitude, remaining within local Ventspils tower control. Given the latency of about 20 minutes, there was always a possibility that these two Mig-23 Flogger-F's could have checked-in with a Ground Controller to go feet wet in pursuit of our Aries-II bird.

I tuned one of my receivers to the GCI frequency for the Ventspils area, and sat on it. Nothing new. Minutes passed, and the Floggers didn't show up on either their local airfield channel, nor on the GCI channel. I put another receiver in general search to cover the customary 30-Mhz segment of the VHF spectrum that these PVO fighters usually used for tactical comms. Time transpired... still nothing. Our two guys covering Soviet Baltic Fleet comms channels were having a field day, however. Walt was also busy with Polish patrol boats, OSA-II's and Tarantul-class anti-shipping combatants, exercising in an area East of Gdynia, in the Gulf of Danzig. The ELINT In-line had plentiful hits on Baltic Fleet Ships' navigation and surveillance radars. The Mission Evaluator, LCDR Thoreau (*dual-hatted as the Mission Commander*), had carefully plotted all of the surface tracks, and his situation board clearly proved that the Soviet ships had formed into two operational groups: a forward group intended for anti-mining, anti-ship, anti-sub, and Air Dominance, and an amphibious landing group following behind carrying the full panoply of armament for setting up a strategic beach-head to spearhead an invasion force (i.e. landing craft, pontoons, tanks, air defense missile batteries, armored personnel carriers, special warfare troops, gun-ship helos, etc.).

These two surface formations had presumably set sail out of the Baltic Fleet port of Baltiysk (*kind of like a Soviet version of Coronado and North Island at the end or our Silver Strand in the San Diego California area*), situated West of Kaliningrad, and overlooking the Gdańsk Bay, in the wee hours of the morning, well in advance of our 05:30 Zulu take-off.

To put it mildly, the entire area north of the city of Gdańsk had turned into a veritable Sharks' Tank. There were probably enough tactical nukes aboard this Soviet Armada to wipe the densely populated Polish coastal cities of Gdańsk and Gdynia off the face of the Earth. DO YOU HEAR ME NOW, PEOPLES OF POLAND? *It would be obvious to even the uninitiated that the Politburo meant business.*

From out of nowhere, the REWSON guy, who had been busily tweaking and audio-visually monitoring his collection of black-boxes installed in the rack directly aft of the "Mission Plot" table (where the Mission Evaluator often sat), yelled out:

"I'm intercepting a DATA LINK signal... NEED IT DF'd... **NOW!**" He passed a note with the microwave frequency jotted onto it directly to the Mission Evaluator, who took it to the BIG LOOK operator and to the ALD-8 Direction Finding system operator.

At the same time, the ELINT In-line announced that there were several Bar-Lock radars active along the coastline between Kaliningrad and Ventspils. It was customary for these long-range early warning radars to be active, along with the complementary more capable Tall-King radars. That is how the Air Defense Zone HQ tracked all of the aircraft activity near the coastline, and the radar info was collated by nodes within the Air Defense Zones, who then transmitted the information to subscribers over Manual Morse and Automatic Morse Air Defense communications channels – exploiting these HF communications channels is how CTR2 Jack Tindall made his money.

"REWSON – your data link signal is bearing 036 degrees, possibly coming from an Air Defense Controller South of Ventspils. What do you think it relates to?" LCDR Thoreau had a hunch, but wanted facts.

"Looks to me like a Ground Controlled Intercept data link. It has all the telemetry data required to relay target range, heading, speed and altitude." REWSON was spinning his recorders on it.

"Mak – anything with those Ventspils FLOGGERS?"

"Negative, Gunner – they've gone quiet... were in local TOL's, last heard 30 minutes ago. I don't have any activity on the GCI channel."

REWSON was biting his tongue, and wringing

his hands. He had something VERY important to say, and this is what he said:

"I think we're up against the new Flogger-G. It's being rolled out as a specialized air-defense interceptor variant. It was developed specifically for the PVO. It's got the improved **High-Lark** radar. Its autopilot includes a new digital computer, which is linked with the Lazur-M datalink. Lazur-M enables automated ground-controlled interception with the GCI controller actually steering the aircraft towards the target. All the pilot has to do is control the engine and use the weapons. CIA tells us that these fighters are solely intended for PVO and will exclusively remain in Soviet service – this one won't be exported."

LCDR Thoreau and Gunner were weighing the gravity of the situation. Two Floggers, possibly this new "Golf" variant might be heading out to intercept us at this very moment, in virtual EMCON (*emission control*), meaning that there would be NO verbal transmissions over the GCI frequencies that I was monitoring.

"Jack – does your tracking show any PVO interceptors coming our way?"

"Negative, Mak – most unusual if there's really a pair of Floggers heading our way." Jack was just as perplexed as the rest of us. Experience told us that Flogger pilots didn't trust anything but their own on-board instruments.

But I DID get a hit on the GCI Fighter-reactor Channel:

"Okhotnik-4, ya Sablya-33 – Lazur vyklyuchayu. Idu na globus." (Hunter-4, this is Sabre-33 – I am turning on the Lazur. I am activating target acquisition).

Now I had LCDR Thoreau's and Gunner Stark's undivided attention – "It's a Ventspils Flogger talking to his GCI controller. He's turning off his "Lazur" system (*contains the data link which enables automated ground-controlled intercept*), and is going to activate its target acquisition radar (*globus*)."

"BIG LOOK (*Aries' powerful Radar Surveillance system*) has a HIGH LARK (*the airborne intercept radar carried by Mig-23 Floggers*) hit, bearing 039, range 31 nautical miles – it's just "lighting-up", painting us, but is NOT in target acquisition mode." LCDR Thoreau was reporting over the ALL-Crew ICS. Reports were flying, and Nav shared our precise location with Pilots, Mission Commander, Communications Evaluator, and the Secure Radio Operator to ensure that the obligatory "Interception" reports that were being drafted were accurate to the nearest 3 nautical miles.

At this point the two Ventspils Floggers were closing in on us ***fast***, flying their interception vector of 217 degrees. From the HIGH LARK intercepts that our BIG LOOK operator was tracking, there were in fact two Floggers, one coming in low, and the other approaching high, with about a 2 NM separation behind the low, inbound Flogger.

This kind of an EMCON Ground Controlled Intercept was unheard of. The common modus operandi was for the GCI controller to take command of the interceptors BEFORE they went feet wet, and to VERBALLY vector them into a position within about 20-to-30 NM from the "bogie" (*aviation slang for target*). Our Secure Comms operator was "GRONKING" the hell out of our receivers while sending out the various mandatory reports, which hindered clean signal intercept, but alerted our PARPRO masters about the impending interception. We remained on track, well over international waters, so ALL WAS WELL, RIGHT? *// for a good read on Soviet Aviation PVO capabilities, pilot training, tactics and interception techniques see this recently declassified CIA report*[97] *//*

[97] https://www.cia.gov/library/readingroom/docs/DOC_0000969825.pdf. Retrieved on Dec. 22, 2016.

"Cockpit – they're coming in from our TWO O'Clock... about 25 miles out. One is coming in low; the other one is about 2 angels higher, a mile or two behind. Let me know when you get a visual on them." LCDR Thoreau had lived through quite a few Baltic intercepts, and knew what the Flogger-F's were capable of, but this was the first time that the improved Flogger-G had jumped him... in EMCON!

"Okhotnik-4, ya Sablya-33 – Tsel' v globuse. Peredo mnoy na 40."

Hunter-4, I'm Sabre-33 – I have the target on radar. He's ahead of me at 40 (*kilometers*).

"Sablya-33, Sablya-37, ya Okhotnik-4 – Opredelite tsel'."

Sabre-33, Sabre-37, I'm Hunter-4 – Identify the target.

"They're on us, Gunner. Flight Leader is 40 klicks out, and the GCI Controller asked the Floggers to ID their target."

Almost simultaneously, the Pilot reported over the All-Crew ICS:

"Bandits (*aviation slang for inbound hostile aircraft*) at our TWO O'Clock, about 15 out, two con-trails (*condensation trails*)"

"Okhotnik-4, ya Sablya-33 – Opredelyayu tsel': ehto Amerikanskiy Morskoy ORION, razvedchik s karlikom na khvoste. On na kurse 35 gradusov, vysota 10,100 ."

As I heard it I wrote it onto the report template, and handed it to Gunner:

"Hunter-4, I'm Sabre-33 – I've identified the target: it's an American Naval ORION, reconnaissance aircraft with an "ELF" on the tail. It's on course 35 degrees, altitude 10,100 meters." Gunner shared it with LCDR Thoreau, who added ELINT information, and walked it up to the cockpit to add the visual sighting information before having Secure Comms send it out.

“GRONNNNNKKKK” – the Sky Pig was rolling in mud. It had intercepted the new Flogger-G in action, on an EMCON intercept profile, employing forward hemisphere look-down / shoot-down tactics over international waters. Drinks would be on Gunner if and when we ever recovered in Germany... West Germany, that is. What a laugh we would have over the Flogger Leader’s description of our Sandeman: an ***ELF!***

You just can’t get no *RESPECT* out of these Godless Soviets! They can’t even tell a swarthy, swash-buckling Sandeman from a court jester!

Sometimes we couldn’t even get no respect from our own top-brass. If I had a dollar for every time the NAVEUR Intel guys said that our Baltic missions compared to a self-licking ice-cream cone, I would be one rich PO2! **Now** these nay-sayers had proof positive that we had drawn the new, improved snake out of its hole, and this time it was no mere Viper. This time we had drawn a Death Adder. The REWSON, and ELINT, and COMINT fingerprinting efforts were all tremendously valuable SCOOPS for the Intel Community.

It wasn’t over by any means... the Floggers closed on us, maneuvering to pull-up alongside on our same course and altitude, each of them taking turns to not jeopardize their mission, which was to remain in position in order to bag their target if and when called upon, and until such time when the Ground Controller called them off.

The guys in the cockpit were giving us updates:

“The low bandit broke off, veering off to our Starboard. The high bandit remains on course, about 2 miles behind its Lead. They’re about 10-12 miles out. Chances are that the high bogey will join us overhead, and assume our course, while the low bogey will simulate an aft “SPLASH” (*aviation slang for downing a bogey with a good missile or gunnery hit*).”

What analytics were coming from the cockpit were based on empirical evidence accumulated over the course of career Baltic missions. Our Pilots’ experiences did not take into account what REWSON suspected (was REWSON the real SANDEMAN?): that the enhanced SKY LARK radars on these death adders consisted of the improved Soviet "Sapfir-23P"[98] radar system, which could be used in conjunction with the gunsight for better look-down/shoot-down capabilities, a system designed to counter the increase in low-level threats like cruise missiles. REWSON knew that the "low bandit" wingman was in fact the observer, for now, and that it was going to be the "high bandit" that would simulate taking us down using his new look-down/shoot-down attack profile. What was evident in this "shake-and-bake" flight-scapade was that in a real-world situation we would probably have dived for the white-caps in a last-ditch effort to save our hides. During one of our recent *bierstube* huddles Walt told of the time last year when an EP-3E Aries-II Mission Commander on this same Baltic PARPRO Track had decided that a pair of inbound SU-15 Flagons under Ground Controlled Intercept from a PVO base near Klaipeda were out for blood. This was on the heels of the aforementioned April 1978 downing of the Korean airliner over

---

[98] http://www.seeninside.net/sapfir23_radar.html. Retrieved on Dec. 22, 2016. Sapfir-23 was designed by a team under Chief Designer Kunyavsky for the MiG-23 fighter. A purely air-to-air design, it was the first radar for a frontal fighter designed to allow BVR engagements. It used semiconductor technology rather than vacuum tubes and a method of external coherence in the mode "SDTs" (moving target selection) to detect aircraft flying below the host aircraft. This had limitations- it could only detect targets in the duration between successive pulses, and had multiple "blind" velocities in multiples of its PRF. It used an analogue AVM-23 computer, and a twist-cassegrain antenna. Some sources indicate that early versions could only detect closing targets. The incorporation of an IRST into the MiG-23 may therefore have been intended give a pursuit engagement capability. Sapfir-23L was the radar of the initial production MiG-23 (1970-71). Sapfir-23D was the full standard radar and the first with limited lookdown capability. Search range was 55km against a closing Tu-16 sized target, 45km against a MiG-21. Tracking range was about 35km. Fitted to production MiG-23M. Lots of problems were encountered in service, as it required expert tuning and high quality maintenance. It wasn't uncommon for the detection range to vary 10 times from one set to the next. Weight about 500kg. **Sapfir-23D-Sh was improved with better discrimination of low flying targets and improved ECCM. Fitted to later production MiG-23M, older MiG-23Ms were upgraded to this standard.**

Murmansk by an SU-15 Flagon.[99] When one of the inbound Flagons "Lit-up" his Twin-Scan radar, illuminating the EP-3E, the Mission Commander ordered the Pilot to take IMMEDIATE Evasive Action with an EMERGENCY DESCENT. All loose gear went flying... unfortunately the Secure Comms Operator was in the "head", depressurizing into the *pissoir* (an empty sono-bouy container that was anchored to the bulkhead by a couple steel fastening bands). Unrestrained and unawares, the pissing O-Brancher was slammed against the cubicle's (*about a 3'x3'x8' space*) ceiling, knocked unconscious, and marinated in urine. It was only after the bird leveled out at Angels 2 from its roller-coaster 29-thousand ft. descent, that someone noticed there was no O-Brancher at Secure Comms to send out the Emergency Action Report... the story became yet another **favorite** in the never-ending compendium of Squadron Legends, and when told well, could be counted on for a free drink from fellow *zoomie* (*air force*) aviators.

"Crew – this is the Mission Commander. Protect your GONADS – I repeat... Protect your GONADS! We are being illuminated with Fire Control Radar in a Track-while-Scan mode from the inbound high-flying bandit." We did what we could, although truth be told, some of the

---

[99] https://www.thisdayinaviation.com/tag/sukhoi-su-15tm/ See 4-20-1978 entry. Retrieved on Dec. 22, 2016. On April 20th 1978 a PVO SU-15 Flagon fired a missile against a straying Korean Boeing 707 airliner that had drifted over Murmansk land-mass while enroute from Anchorage to Paris. The missile damaged the 707's tail and wing, sending lethal shrapnel into the passenger cabin. Two passengers were killed. The Korean pilot testified that when he realized that he was being pursued by interceptor-fighters he followed the standard ICAO procedures of lowering his landing gear, flashing navigational lights off and on, and reducing speed, to signal his willingness to follow the interceptors in for a landing, all to no avail. After being hit by the missile detonation, the 707 went into a controlled emergency descent through the clouds, losing the pursuing Flagons. The Korean pilot flew at low altitude over snow covered terrain for over an hour in search for a suitable landing strip, and finally, at nightfall, and after several aborted attempts due to obstacles near short runways, he made a smooth landing on a frozen lake bed near the city of Kem.

blokes were okay with sterilization… might help them score more easily when on liberty. I reached for one of my metal "plotting" clip-boards, shielded the jewels, and kept at it, gisting away. The ones really at risk in the naval profession were my nuke-submarine compadres… couldn't help but feel for them, and be glad that I was airborne.

"Okhotnik-4, ya Sablya-33 – Razreshite prakticheskiy ZAPUSK. Tsel' v Globuse."

My report: "Hunter-4, this is Sabre-33 – Request permission for a simulated LAUNCH. Target is on (my) radar.

"RAZRESHAYU!"

Permission granted!

"Chto?"

What?

"***RAZ-RE-SHA-YU***!"

PER-MI-SSION GRAN-TED!

"*Ponyatno. Lazur vklyucheno*."

Understood. Lazur turned-on.

That was the last we heard from Lead Interceptor Sable-33 as he proceeded to simulate shooting us down with an air-to-air missile. The simulated launch had possibly been handed over to the GCI commander, so that the data link was uploading target engagement/destruction commands directly to the Flogger-G's Sapfir-23P weapons control system. In REWSON's mind, this marked the beginning of feasible drone-interceptors. What was once science fiction was now reality, and it was conceivable that Unmanned Aerial Vehicles (UAVs) could *robotically* conduct air-to-air warfare over data links with a GCI command post in charge of the entire operation, from take-off, to shoot-down, to landing.

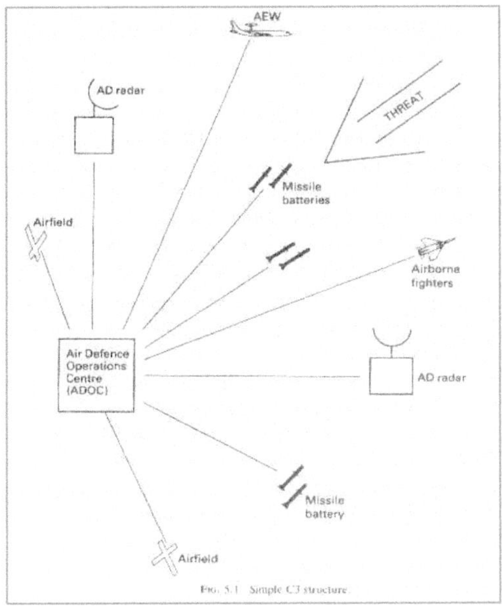

Fig. 5.1 Simple C3 structure

Not to be outdone, the Flogger Wing-man jumped into action.

"Okhotnik-4, ya Sablya-37 – Uspeshnyy ZAPUSK. Tsel' unichtozheny."

My report: "Hunter-4, this is Sabre-37 – Successful LAUNCH. Target Destroyed."

This was yet another simulated SPLASH, only this time the Flogger wingman had conducted a rear-hemisphere intercept. Since the Flogger never indicated in comms that he had acquired "LOCK-ON"[100], nor did the ELINT intercept indicate that the fighter-interceptor had at any point switched his High-Lark radar from "Track-while-Scan" into "LOCK-ON" mode, it was up to the Mission Commander to interpret whether or not this was an Incident at Sea, which required yet another type of report. Best to let the EUCOM staff figure it out… we had plenty on our hands, and the Soviet Naval Amphibious Assault on Poland was now progressing into Phase-II:

---

[100] http://www.radartutorial.eu/02.basics/Fire-control%20radar.en.html. Retrieved on Dec. 22. Fire-control radars operate in three different phases: a. Designation or vectoring phase: The fire-control radar must be directed to the general location of the target due to the radar's narrow beam width. This phase is also called "lighting up". It ends when lock-on is acquired. b. Acquisition phase: The fire-control radar switches to the acquisition phase of operation once the radar is in the general vicinity of the target. During this phase, the radar system searches in the designated area in a predetermined search pattern until the target is located or re-designated. This phase terminates when a weapon is launched. c. Tracking phase: The fire-control radar enters into the track phase when the target is located. The radar system locks onto the target during this phase. This phase ends when the target is destroyed.

anti-shipping operations and Air Dominance to clear the way for a successful amphibious landing.

Now the Low Bandit Flogger (i.e. the wingman) pulled up along our Port (*left*) side, looking directly at our First Pilot, and at the Secure Comms Operator. There was a flurry of picture-taking going on in the cockpit. Secure Comms, whose Posit was on the Port side, just aft of the flight deck, and had the most beautiful *large* fish-eye window on the entire aircraft, came onto the All-Crew ICS:

"This is Secure Comms – I'm looking directly at a Flogger and by God, he's looking directly back at me. His wings are tilted forward, flaps out... intends to stay with us. DOES ANYBODY HAVE A PLAYBOY CENTERFOLD? BRING IT TO ME ASAP!"

Wouldn't you know it, that one of Slap-shot's post-repast activities was to flip though the recent porn trove he had gathered at the Stars and Stripes bookstore?

In seconds flat Slap-shot was crowding Secure Comms' fish-eye window, with the Playmate of the Month, Miss September 1979[101] Vicky McCarty flashing her birthday suit endowments back at "Sablya-37", the wingman interceptor. "Sablya-33", the Lead interceptor, had looped aft of us, then overtaken our aircraft from below, and now took station about a mile ahead, at our ONE O'Clock – no eye-candy for the top dog. Both Floggers had trouble with low speeds... our pilot throttled back and reduced our airspeed. Within minutes they were gone.

"Okhotnik-4, ya Sablya-33 – Vozvrashchaemsya domoy."

Hunter-4, I'm Sabre-33 – we're returning home.

---

[101] http://www.playboy.com/vicki-mccarty. Retrieved on Dec. 22, 2016. Vicky's least favorite phrase: "What's your sign?"

LCDR Thoreau conferred with the Pilot, and decided that now that the Bandits were gone, it was a good time to reverse course, and head back to establish an orbit North of Kaliningrad, to collect in detail on the Amphibious Landing operations.

Throughout the remainder of the mission we would continue to haul-in SIGINT from not only the military ships and aircraft operating in Baltic waters, but also from the military naval, army, and air-defense bases in an area stretching from Ventspils, Latvia to Peenemünde[102], East Germany. Upon our recovery at Echterdingen some stayed with the aircraft to ready it for tomorrow's mission. Others of us boarded shuttle buses to head back to EUCOM Vaihingen and Patch Barracks.

"Gunner – this Post-Mission Report should only take about two hours… we might have to delay tomorrow's show time to get our 8 hours of crew rest."

"The hell with crew rest – that only applies to the Pilots." We all knew the facts: no rest for the devil. "Show time for the shuttle is as scheduled. It's a 05:30 show for a 07:30 go."

With that settled, it was high time for some levity;

"So, LCDR Thoreau – how does it feel to be part of the "Elvin" Squadron?"

---

[102] Klee, Ernst; Merk, Otto (1963, English translation 1965). The Birth of the Missile: The Secrets of Peenemünde'. Hamburg: Gerhard Stalling Verlag. Peenemünde was the site of the Nazi Army Research Center run by Werner Von Braun from 1937 to 1945 that designed the V-2 guided missiles used in WW-II. Von Braun also directed research for developing a more capable V-3 missile intended for use against the USA. The ARC also developed the Wasserfall (35 Peenemünde trial firings), Schmetterling, Rheintochter, Taifun, and Enzian missiles. The Peenemünde establishment is credited with the development of the first closed-circuit television system in the world, installed at Test Stand VII to monitor rocket launches.

Wolfcamp

Military operations, whether large or small, take on code names to ensure that only those "in the know" get access to all of the ongoing developments through compartmented intelligence channels. I've mentioned Operation Ivy Bells, in which the Navy, the CIA, and the NSA joined forces to tap Soviet underwater communications trunk-lines in the Sea of Okhotsk. That was a major production, employing countless resources on a National scale. There were myriad small *Cold War* operations that also tapped resources from the DoD Services (Army, Navy, Air Force, Marine Corps), and the Intel Community at large.

One of these small operations was code-named "Wolfcamp", an initiative that had sprung from indications in early 1979 that the Soviet Union was once again actively employing a Brigade of stand-alone ground forces in the form of a newly forward-deployed Red-Army brigade. Congress and the American power-brokers knew that if this were the case, the Soviet Union would be, once again, violating the Monroe Doctrine[103], an untenable situation for our homeland security, given that the last time that the Soviets had violated our Monroe Doctrine was during the Cuban Missile Crisis, a chapter in history that could well have resulted in nuclear holocaust for humanity.

News that this Soviet brigade was operating in Cuba coincided with U.S. State Department initiatives to negotiate more serious Strategic Arms limitations with Brezhnev's Kremlin. The era of "Détente" that Henry Kissinger had paved the way toward had given rise to Cyrus Vance's U.S.-Soviet Arms Control initiatives in the form of meticulous SALT-II negotiations.

---

[103] Monroe Doctrine: a principle of US policy, originated by President James Monroe in 1823, that any intervention by external powers in the politics of the Americas is a potentially hostile act against the US.

Figure 1. Soviet Military Facilities in Cuba, 1962

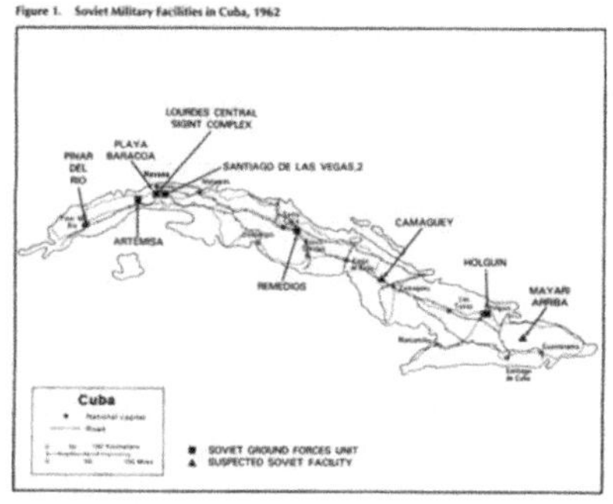

But the spirit of Détente was highly contested by the then National Security Advisor to President Jimmy Carter – Dr. Zbigniew Brzezinski[104], who not only pressed vociferously for Human Rights throughout Soviet-dominated Eastern Europe (Dr. Brzezinski was known to confer with Pope Karol Wojtyla, now a.k.a. Saint John-Paul II, on strategies to liberate Poland from the Soviet bear-hug), but who pointed out that Détente was a mere illusion, one that the Soviet Union was using to mask its true nature of imperial expansionism. As a matter of fact, the Soviet Union would soon engage in its own Viet Nam by invading Afghanistan in December of 1979[105]. So the

---

[104]https://repository.library.georgetown.edu/bitstream/handle/10822/553133/sextonMaryDuBois.pdf?sequence=1 Retrieved Dec. 22, 2016. By 1978, Brzezinski and Vance were more and more at odds over the direction of Carter's foreign policy. Vance sought to continue the style of détente engineered by Nixon-Kissinger, with a focus on arms control. Brzezinski believed that détente emboldened the Soviets in Angola and the Middle East, and so he argued for increased military strength and an emphasis on human rights. Vance, the State Department, and the media criticized Brzezinski publicly as seeking to revive the Cold War. Brzezinski anticipated the Soviet invasion, and, with the support of Saudi Arabia, Pakistan, and the People's Republic of China, he created a strategy to undermine the Soviet presence. Using this atmosphere of insecurity, Brzezinski led the United States toward a new arms buildup and the development of the Rapid Deployment Forces.

[105] https://history.state.gov/milestones/1977-1980/soviet-invasion-afghanistan. Retrieved on Dec. 22, 2016. At the end of December 1979, the Soviet Union sent thousands of troops into Afghanistan and immediately assumed complete military and political control of Kabul and large portions of the country. This event began a brutal, decade-long attempt by Moscow to subdue the Afghan civil war and maintain a friendly and socialist government on its border. It was a watershed event of the Cold War, marking the only time the Soviet Union invaded a country outside the Eastern Bloc—a strategic decision met by nearly worldwide condemnation. While the massive, lightning-fast military maneuvers and brazenness of Soviet political objectives constituted an "invasion" of Afghanistan, the word "intervention" more accurately describes these events as the culmination of growing Soviet domination going back to 1973. Undoubtedly,

table was set for ensuring that any limelighting of the Soviet activities in Cuba would antagonize U.S. State Department SALT-II negotiations. President Carter was pulled in different directions by his National Security Advisor (Brzezinski) and his Secretary of State (Vance) well before the Iran Hostage Crisis[106] reached its apogee.

President Carter's 1979 predicament must have been the source of inspiration for singer Gerry Rafferty's "Stuck in the Middle with You":

Trying to make some sense of it all,
But I can see that it makes no sense at all,
Is it cool to go to sleep on the floor,
'Cause I don't think that I can take anymore
Clowns to the left of me, jokers to the right,
Here I am, stuck in the middle with you.

The "YOU" alludes to the noble "We, the American People" who had elected President Jimmy Carter to weather the storm for all of us.

Political pressures triggered President Carter's speech on "Soviet Troops in Cuba", published by the State Department as Release 92 on October 1, 1979. President Carter announced new measures to deal with this "Soviet Combat Brigade". The new measures included increased surveillance over Cuba, and over joint Cuban-Soviet military activities around the world. This was the genesis of Operation "Wolfcamp".

---

leaders in the Kremlin had hoped that a rapid and complete military takeover would secure Afghanistan's place as an exemplar of the Brezhnev Doctrine, which held that once a country became socialist Moscow would never permit it to return to the capitalist camp. The United States and its European allies, guided by their own doctrine of containment, sharply criticized the Soviet move into Afghanistan and devised numerous measures to compel Moscow to withdraw.

[106] "The History Guy: Iran-U.S. Hostage Crisis (1979–1981)". Historyguy.com. Retrieved Dec. 22, 2016. The Iran hostage crisis was a diplomatic crisis between Iran and the United States. 52 American diplomats and citizens were held hostage for 444 days (November 4, 1979, to January 20, 1981) after a group of Iranian students belonging to the Muslim Student Followers of the Imam's Line, who supported the Iranian Revolution, took over the U.S. Embassy in Tehran.

"Look, Petty Officer Mak, I know that you're a short-timer with only months left onboard, but this could be one of your best deployments yet. The operation is being run out of the NSA, and they are looking for airborne linguists that are dual-qualified in Russian and Spanish. Your name came up. Believe me, this is **big**. It has nothing to do with our new-fangled operations in the Caribbean, that piss-ant stuff to see what kinds of weapons are being delivered to the Sandinistas."

"Yeah, but Chief – the schedule has me pulling one last Whale Det onboard the *Forrest Fire* in December, maybe even missing Christmas with my family."

"I can guaran-fu..ing-tee you that this one is a short deployment. Do good work, get some recognition from the head honcho, and you can skip the Forrestal Det. We'll get one of the bachelors around here to cover it. This mission is designed to suck up all of the microwave back-scatter from the General Headquarters command and control links coming into and out of Havana. The NSA has been tasked to prove that there's a Soviet combat brigade operating in Cuba, and to get the scoop on what it is that they're up to. If you're selected, you'll get orders to attach to the Project Director of **Operation Wolfcamp** as the Lead Linguist in charge. Here's the tasker."

Chief Light showed me a message from the CNO N2 (*Office of Naval Intelligence*) that briefly described tasking for an "airborne platform" to conduct reconnaissance operations in international waters North and East of Havana in support of Operation Wolfcamp." It referenced other messages from the Office of the Secretary of Defense… messages that Rota ditch-diggers did not have access to.

"Ok, Chief – put my name in the hat. What kind of aircraft?"

"Don't know – but given the microwave back-scatter requirement, it has to be something slow. Maybe a helo."

Now I had nothing against helos, but my only experience in helos was a water survival training exercise where a bunch of us flyers were told to jump out of a perfectly good helo into the waters of the Bay of Cadiz about 30 feet below. We did this in the early a.m., middle of July, while suited up in full flight gear. The purpose was trivial - only to inflate our life vests, wait to be picked-up in a horse-collar rig, and get hoisted back up into the helo. Talk about *GOOD TIMES*! Full flight gear (helmet, flight suit, flight boots, survival vest, gloves, etc.) a good swimmer does not make. My take is that the exercise was more about training the Search and Rescue (*SAR*) helo crew than it was about building any of us "downed aircraft" pogues'[107] survival skills.

I had also almost hitched a ride on a Navy helo going from Moron Air Base near Seville, to Rota when my Skywarrior had broken down on its way home, and had to divert to Moron, which was only about 80 miles from Rota. I remember yelling across the tarmac to a nearby Navy helo pilot to ask where he was headed. When he said "Rota!", I threw my flight bag over my shoulder, and raced out to the already rotating bird. Much to my surprise, when I was about 50 yards from the chopper, an Air Force Security Truck swooped in, stopped directly ahead of me, and two armed Military Police jumped out, weapons ready.

"Whoa, Ace – you got this wrong… I'm hitching a ride with a fellow Navy aviator to get back to my homeplate in Rota."

"Come with us – you're being detained!"

The story might seem preposterous, especially since the Security Unit had seen me get off the EA-3B Skywarrior, sling my flight bag over my shoulder, and jog toward the helo. In anyone's book, by anyone's standards, this could be seen as an "official" hitch-hike.

---

[107] **Pogue** is pejorative military slang for non-combat, staff, and other rear-echelon or support units. "**Pogue**" frequently applies to those who do not have to undergo the risk and stresses of combat as the infantry does.

I was apprehended, locked in the rear cab, and taken to the duty officer where I was told to present my ID card and credentials, which I did – GI mug-shot, Spanish Status-of-Forces yellow ID, and my set of orders. None of the guys from my crew had seen the Security Unit haul me away. They probably assumed that I had caught a ride with the helo, so happy trails. My bubbas were probably now checking-in at some hotel out in town, near Seville, home of the *Sevillanas*[108]. Today's cell phone technology would have helped me nip this fiasco at the bud.

"Says here that you report to NAVCOMMSTA Rota, Naval Security Group Department."

"That's right, Master Sergeant. What's the charge?"

"I'll ask the questions.... Petty Officer... *Mar-kins-key*. Do you have a phone number for your Commanding Officer?"

"The Command Quarter Deck phone number is on the copy of my orders... on your desk."

"Ok – what's your Commanding Officer's name?"

"Chief Light." These zoomies were clueless about Navy ranks.

He dialed... someone answered... he asked for Chief Light.

"Chief Light – this is Master Sergeant Klusterfill at Moron Air Base Flight Line Security Post One. One of your sailors is being detained for having violated a Security Line... his name is *Mar-kin-skey*."

"You detained who? Petty Officer Mak? He needs to get back to home plate pronto... Don't hold him up!"

---

[108] Best source is Spanish: https://es.wikipedia.org/wiki/Sevillanas. Retrieved Dec. 22, 2016.

MSgt. Klusterfill hung up. It was plain to me that he expected this outcome, but got some kind of ape-ish joy in having flexed the security-wonk muscle.

"Ok – Petty Officer *Mark-rinse-key,* here's your ID cards and orders."

"Where's the "Security Line", Master Sergeant?"

"It's an imaginary circle described by a 200-foot radius around our RC-135 Rivet Joint aircraft, the one parked on the other side of the helo you were wanting to catch."

Damn, I thought to myself... this is the stuff of Joseph Heller[109] novels... come to think of it, Heller's "Catch-22" is all about zoomies in WW-II Italy. I wholeheartedly agree with Joe's point of view, in that these zoomies can really put a fine point on living in a "theater of the absurd."

Several days passed from the time when Chief Light and I had the conversation about Operation Wolfcamp. On day three there was movement.

"Mak – here's the count: there are about half a dozen professionalized active duty Russian-Spanish linguists in the DoD, but you're the only one who's Airborne qualified. They're expecting you to report to Naval Air Station Key West[110] on the Boca Chica Key within 48 hours."

---

[109] Joseph Heller – American Novelist 1923-1999, the phrase "Catch-22" has since entered the English language, referring to a type of unsolvable logic puzzle sometimes called a double bind. According to the novel, people who were crazy were not obliged to fly missions, but anyone who applied to stop flying was showing a rational concern for his safety and was, therefore, sane and had to fly. To quote Heller's protagonist in Catch-22, Capt. John Yossarian: **"He was going to live forever, or die in the attempt."**

[110] http://www.globalsecurity.org/military/facility/moa-keywest.htm. Retrieved on Dec. 22, 2016.

"I gave Victoria and the kids a heads-up over the week-end, so now it's a matter of packing my flight gear and a pair of jeans. I'll head home, but first, what kind of tech kit should I put together?"

"No need – there's a tech dump for you at the Fleet Electronic Warfare Support Group (FEWSG) Intel Center. Your orders, a cash advance, and your plane tickets will be waiting for you at the Quarterdeck. Good luck, and have a Mojito or two at Sloppy Joe's for me."

I had no idea what he was talking about, but I made a mental note of it... might serve me well if and when the Director of Operation Wolfcamp gave me some time off. Half the fun in globe-trotting is getting to see what it is that makes the locals tick.

My itinerary was hurried – a military hop to Norfolk, and from there a military C-12 (*Beechcraft "Huron"*) into NAS Key West, where one of the FEWSG duty drivers met me to get me over to the Bachelor Enlisted Quarters. The front desk Petty Officer said there were no rooms available:

"Sorry, buddy. The transient rooms are all taken. We've had bunches of aircraft maintenance detachments check in this week. Most of 'em Army. Look around, and you'll see that a boat-full of *doggies*[111]"have taken over our BEQ."

"Any hotel nearby with base shuttle service?"

"'Fraid not." He stamped my orders *"Government Quarters NOT Available"*

---

[111] Origin of the term "doggies" for Army troops – *The Reader's Digest* described one WWII GI as saying, quote: "The Army treats you like a dog. You have to wear dog tags, sleep in a pup tent, eat dog food-and they yell at you when you growl."

I looked my orders over, and the "Authorized Rental Car" box wasn't checked... no rental car, but taxis were authorized – a frugal sponsor, this NSA.

I got a ride to the FEWSG Quarterdeck.

"Petty Officer Makfinsky, reporting for duty. I just came in from NAVCOMMSTA Rota, NAVSECGRU Division for a project that's being coordinated out of your organization."

"Welcome aboard, Petty Officer Makfinsky. I'm Senior Chief Castro. We've been expecting you – although the entire team is not assembled yet. The project Lead arrived yesterday, as did his two Huey choppers (i.e. *UH-1H Bell Helicopters, affectionately called Baby Hueys by Army Rangers*) from the 1st Cavalry. The birds have been here since Monday last week. Wild bunch of dudes... they have Warrant Officer pilots and navigators, and these Warrants ain't got religion, believe me."

I didn't have to salute the Senior Chief, but I did, given his positional authority as Officer of the Deck. He boasted an Intelligence Specialist (IS) Quill and Magnifying Glass rating insignia on his rank patch

– honoring the time-tested quality of IS's in imagery reporting, although their skillsets extended far beyond imagery analysis.

"Senior – here's a copy of my orders. I'll need to contact homeplate in Rota to close loop and let them know that I'm here."

"No sweat – here's an AUTOVON (*automatic voice network – a world-wide military phone system*). Dial 9 for a tone, and then 97 for Europe... I take it you have your NAVSTA Rota number."

"Thanks, Senior Castro. Sure do."

"Chief Light – I've checked in with the *FEW-SIG*, just got my orders stamped 20:32 Zulu. All is well, and I'll hook-up with the Director tomorrow."

Chief Light acknowledged and hung up. AUTOVON was not used for much more than "ET-phone home" status reports. Ever since the CIA had experienced several damaging exposures from insider espionage (i.e. Edwin Moore II in the early 70's, and David Henry Barnett in the late 70's[112]), our counter-intelligence (*CI*) and operational security (*OPSEC*) briefings continuously stressed the importance of minimizing conversations over open lines.

Project WOLFCAMP details have been declassified but remain sensitive. We proved the presence of a Soviet Red Army authority in Cuba communicating over its own command and control network. Should methods and sources become declassified on the heels of a major change in Cuban government, I will be able to proceed with this story.

---

[112] http://content.time.com/time/magazine/article/0,9171,924498-1,00.html. Retrieved on Dec. 22, 2017. David Henry Barnett was a CIA officer who was convicted of espionage for the Soviet Union in 1980, becoming only the second CIA officer to be convicted after Edwin Moore II, a retired CIA employee who was arrested by the FBI in 1976 after attempting to sell classified documents to Soviet officials.

All Good Things...

You know the saying. As exciting as Airborne Cryptology had proven to be, it was time for Victoria and me to move on. We were a family of five now that Paul and Daniel had sprung to life at the Naval Hospital in Rota, Spain. The obstetricians who oversaw their births, Dr. Lyons and Dr. Seeds, were two real personalities, genuinely fascinated by the miracles of life that they orchestrated in their O-R's. One of them was, matter-of-factly, an excellent tennis player who I routinely encountered on the tennis courts behind the Bachelor Officers' Quarters (BOQ). Not only was he a tough, take-no-prisoners kind of a tennis player, he also enjoyed putting money on his game, and was one of the movers and shakers in the Rota Tennis Association – the RTA, an athletic organization that included men, women, youth, and children from all cuts of the Rota population: Spanish and American; military and civilian; young and old; retired and still employed; bankers and lawyers; doctors and nurses; aviators and ship-drivers; sailors and marines; teachers and students... the list goes on, since the population onboard the Rota Naval base was probably circa twelve thousand in the late '70's.

These were the days of American tennis giants[113]: Jimmy Connors, Arthur Ashe, Pete Sampras, John McEnroe, and Chrissie Evert, to name a few. Their achievements in international competitions were an inspiration to an entire generation of Americans, and our national enthusiasm for the sport was evident in Rota Naval Base social life. Just as important were Spain's tremendous credentials in the tennis world: Manolo Santana and Manuel Orantes had brought fame and fortune to the Spanish Tennis Federation through their Davis Cup, U.S. Open, and

[113] http://bleacherreport.com/articles/1379435-ranking-the-greatest-us-tennis-players-by-decade. Retrieved Jan. 2, 2017.

Wimbledon successes, so that the Rota Tennis Association had a mix of about 50-50 Spanish-American membership. One of the most notable events that Dr. Lyons directed, as an associate chair of the RTA, was a Calcutta Tournament[114] - a competition that brought all players together on an equal footing into the most massive hurrah that the Association had ever known. Much of the planning was done by veteran members in a bar-restaurant located in El Puerto de Santa Maria, called "El Resbaladero" (*The Slippery Place*), a quaint, somewhat historical locale.

---

[114] http://thepaseoclub.com/wp-content/uploads/2014/03/What-is-a-Calcutta-Tennis-Tournament.pdf. Retrieved Jan. 2, 2017. A Calcutta tournament is a tournament where teams compete on a handicap system. The better the team, the lower handicap they receive. The main idea is that every team has an even chance to win the match and tournament, no matter what their level of play is. They play for a piece of the prize fund, which is determined by the total amount of money bet on all the teams. Format of Play: Doubles Teams of any combination may enter. If you don't have a partner, organizer provides. •Each match is one set. The first team to reach 41 points is the winner. •Each team is given a handicap, the better the skill level of the team, the fewer the points they receive on their handicap. •Each match requires a score keeper to keep track of the score. •Everyone that plays in Calcutta is required to score at least 1 match and it is very easy to do •When 2 teams play, the difference between the two team's handicaps is how the match is started. •If a team loses in the first round it plays, it goes into a separate consolation tournament. How does the Betting Work? •On Friday night, we serve dinner and have the bar available to purchase drinks. •Each team is auctioned off to the highest bidder. •There are 2 ways to make money in a Calcutta: 1) Play well enough to finish in a spot that earns money and make sure you "buy back" your team. 2) Buy another team that does well enough to earn money. •The 2 players on the team have the right to "buy back" their team up to 50%. If a team sells for $100, then each player may pay up to $25 and that entitles them to 25% of whatever their team wins. •Members generally decide how much they are going to bet and pool their money together with other members to form a "syndicate". •After each team is auctioned off, the money is collected. •The draw is made and is called a "blind draw" as the teams are drawn out of a hat and entered from top to bottom on the draw sheet. •The total money collected is divided into the main draw (80% of the money) and the consolation draw (20% of the money). •This money is distributed to at least the semi-finalists of the main draw (usually the quarter finalists will also earn some money), and the finalists of the consolation draw. It's FUN.

During the 50's and 60's it had served as the fish distributorship for the city... hence the allusion to slippery-slidy floors. There the RTA vets, over glasses of Fino Sherry, and "tapas" food morsels, concocted what was to be the most amusing tennis event that Victoria and I have ever participated in. Once the rules of play, team formation, and betting were laid down, the socialization phase kicked-in.

"El Resbaladero" Restaurant, circa 1978

On the Friday night before the Spring Fling Saturday tournament, we all (players and spouses) went to dinner at a newly-opened two-story restaurant out in town called "El Faisan Dorado" (*The Golden Pheasant*), known for its excellent food and abundantly stocked wine cellar. Another reason why *The Pheasant* was chosen was because its upstairs dining area could seat well over a hundred persons while at the same time maintain an open bar at one end of the salon, and to be sure, these athletes were ready to party before "betting down". At the appropriate point, when most revelers were digging into their dessert and some were well into after-dinner drinks, Commander Lyons rang his water glass with silverware.

"Hear, Hear RTA. Here are the rules: We need an unbiased hand to run our "blind draw" for the doubles pairings. As the names are drawn from this Sangria pitcher for team pairings, they will be entered from top to bottom on the draw sheet. Teams will consist of a doubles pair from the A flight, a doubles pair from the B flight and a doubles pair from the C flight, so six players per Team. We will then auction off each team to the highest bidder."

An unbiased spouse unfolded name-slips as she drew them from the pitcher. Soon the Tournament Roster of doubles teams was complete, and now the bidding frenzy began. Since not all participants fully understood the bidding wars, Commander Lyons would frequently halt the process, explain the proceedings (with an interpreter alongside for the benefit of the Spanish RTA members), and when all were onboard, resume the party. The grand thing about it is that the party had to be called to order frequently, as empty dessert plates were retrieved by bemused waiters, and after-dinner libations loosened up the billfolds. The auction was a resounding success, and the party then dispersed to the four winds, (sure, some were already four sheets to the wind), groups having formed around the various teams, to party-on at downtown locations, or at players' homes. Victoria tells me we went to one of the highest bidder's homes in the Vistahermosa subdivision (not too far from *El Faisan*), but I honestly have no recollection. Could be that the couple hosting us had bought the team I was on, and wanted to keep a close eye on me, lest I be too hung-over to score a victory on Saturday.

There are 2 ways to make money in a Calcutta: 1) Play well enough to finish in a spot that earns money and make sure you "buy back" your team. 2) Buy another team that does well enough to earn money. When a team is purchased (i.e. a final, top bid has been declared by the auctioneer), the 6 players on the team have the right to "buy back" up to 50% of their team. So if a team sells for $120, then each player may pay up to $10 and that entitles them to one sixth of whatever their team wins.

To make a long story short, on Saturday the Tournament convened at the Vistahermosa Tennis Club. Most teams arrived promptly, but some were definitely behind the power curve, as players nursed a bent wing with Bloody Maries and such. What was certain was that the younger the player, the better his or her chances of "winging it", so to speak.

Tennis attire in the 70's was all about *all-white*, and the only fashion statements to be made had to do with an abandoned sculpting of the brawn for men, and an alluring flashing of the curves and struts for women (*remember Anna White's most entertaining full leotard suit at the 1985 Wimbledon Open*). In addition to cunning forehand and backhand strokes, ball-crushing serves-and-volleys, and dashing defensive "gets", one weapon that proved strategically significant in this Spring tourney was the "cleavage drop-shot" (or better yet, the "dropped-cleavage shot"), which seriously hindered many stalwart opponents through surprise and distraction. In spite of the many distractions and tactical contretemps between my partner and me, we managed to place second, and did our part to move the Team into prize dough. In the aggregate our Team placed Second (*by adding the total number of games that our A, and B, and C flight doubles teams had scored, and comparing to the rest of the field's aggregated score*). The "syndicate" that had bought into our team was also rewarded.

When the matches were over, the event continued at the Vista Hermosa Casa Grande[115] where players emerged spiffed-up from their respective

locker-rooms to attend the Awards Banquet, again hosted by Commander Lyons and his RTA entourage. It's safe to say that most, if not all of the prize money ended up in the bartender's cash register.

Re-"Upping"

It was tough to leave all this behind, but my re-enlistment (*to re-enlist is to re-up in Navy slang*) contract was not only about a decent pay bonus (CTI Russian linguists were much in demand in 1980), but more so about moving up the ladder as a consecrated Soviet VMF (*Voenno-Morskoy Flot – or Military Naval Fleet*) cryptologic analyst. For this my detailer – CTICS Jim Yeargain (a.k.a. *Silver Fox* after his mane of silver tinsel), an Arabic linguist of renown, decided that my best career move would be to take on a Baltic Fleet Analyst job at the United Kingdom's General Communications Headquarters (GCHQ), which is the UK's counterpart to our National Security Agency (NSA). Once there I would learn from masterful Royal Navy Warrant Officers who had specialized in this very field over the course of their lengthy professions. Also in the wings was my application for Officer Candidate School (OCS), which my Division Officer, Lieutenant Commander Paul E. Christensen had generously endorsed and forwarded to the Bureau of Personnel.

---

115 La Casa Grande de Vistahermosa now serves as the Golf and Racquet Club's clubhouse. It was built by the Osborne family in the early 1900's as a summer retreat from the overbearing Julys and Augusts in Seville. http://gestion.vistahermosaclubdegolf.com/page/aboutus. Retrieved on Jan 2, 2017.

My re-enlistment ceremony was held on an EP-3E Aries-II DEEPWELL bird, perhaps to underscore my roots as a COMINT Skywarrior, destined to return to the sky one day.

We shipped our car, put our household goods in storage, threw one *hell-a-cious* Farewell party at the Hotel Playa de la Luz[116], on Rota's North Beach, and caught a military flight from Rota Naval Station to Mildenhall Air Base near London, UK. From there we were shuttled to Cheltenham, a bucolic rural town Northwest of London, in the rolling hills of the Cotswolds (*wold* = hillside).

My new Command, the Special U.S. Liaison Office – GCHQ (SUSLO)[117], was run by a SIGINT-savvy Lieutenant Commander who was filling a post that was earmarked for a full Commander. SUSLO "sponsored" my family and me into temporary lodging at the Cotswold Grange Hotel in Cheltenham. The contrast against life in Rota, Spain was stark: from sunshine, aviation, beaches, family outings, tennis and partying, to rain, tight living quarters, office bureaucracy, lack of activities for Victoria, and lack of schools for the boys. Career moves can be jarring. Gone, for the time being, was "La Buena Vida NAVY".

GCHQ Cheltenham

When you hold Top Secret, compartmented security clearances (*"tickets"*) for access to, and the production of sensitive classified information, your whereabouts must be accounted for so that there is no

---

[116] http://www.hace.es/en/Hotels-Cadiz-Spain/Rota/Hotel-Playa-de-la-Luz. Retrieved on Jan 2, 2017. The Hotel Playa de la Luz is located between sand dunes, a pine forest and the Atlantic Ocean. The waves break against the hotel's facade and you can hear and smell the proximity of the ocean from every corner of the hotel.

[117] The SUSLO relationship was established post-WW-II: 6 Apr. 1954 USCIB Directive 12 regularizes the SUSLO system for COMINT liaison with UKUSA countries.

48-hour void - **ever**. Failure to comply with this standard procedure would result in an SAER (Security Access Eligibility Report) against your clearance, meaning that an investigation would follow, and your access would be suspended until the investigation concluded. Other factors that triggered SAERs were security violations (failure to handle classified information per regulations); risky lifestyle changes (sexual activity with foreign nationals, or incriminating and illicit sexual activity); unaccounted for wealth; erratic behavior; and consultation with mental health professionals.

As such, the first order of business for me at the GCHQ was to suit up in my "civvies" (for which I had received a special clothing allowance, since military uniforms were not to be used by SUSLO personnel), and check-in with the SUSLO Special Security Officer (SSO). From there I was to be indoctrinated into the complex personnel security procedures that had to be religiously followed at the GCHQ, which was in the midst of a Ministry of Defense (MoD), and possibly even a Parliamentary investigation over the ***Geoffrey Prime*** espionage case under the Official Secrets Act[118].

After signing a batch of papers and obtaining a serialized number badge from my SSO, I followed the check-in process, and visited the UK GCHQ Special Security Office. Not much new – more physical security briefings, explanations as to how the various buildings had different access requirements, etc. I asked whether I would have to sign for any cypher combinations to vault doors (such was the practice at

[118] I arrived at GCHQ in the summer of 1980, when it was in the midst of an espionage investigation on Geoffrey Prime, who turned out to be a British version of USA's John Walker. Hansard: May 1983 Geoffrey Arthur Prime Security Commission Report". Hansard. 12 May 1983 http://hansard.millbanksystems.com/written_answers/1983/may/12/geoffrey-arthur-prime-security#S5LV0442P0_19830512_LWA_13. Retrieved Jan. 2, 2017.

the Naval Security Group Department, at Building 533 in Rota Spain, and presumably throughout the entire United States Intelligence Community), and was surprised by the answer:

"I'm getting you an appointment to meet with Mr. Lantham and Mr. Spivey, the two security officers at the front gate." The young lady never touched on my cypher combination question.

When she got off the phone, I was told to meet with Mr. Lantham and Mr. Spivey at the main entrance; within the next ten minutes... they were expecting me.

I climbed some steps to the main entrance, flashed my serialized badge to a middle-aged fellow sporting a blazer and dark necktie who sat alongside one of the many turnstiles, and mentioned to him that I needed to meet with Mr. Lantham and Mr. Spivey.

"And what is your name, Sir?"

"Michael Makfinsky."

"And your affiliation, may I ask?"

"Suhs-Low" – I replied.

"Ah – yes. I'm Tom Spivey, and my colleague, Terry Lantham at the desk over there will need to have a word with you."

I pushed through the turnstile, and met with Lantham at his desk, which was out in the open, dominating the lobby beyond the row of turnstiles.

"Welcome – please have a seat."

Mr. Terry Lantham had all the composure of a Formula-1 race car stable manager. I could easily picture this gray, swarthy figure managing Team Brabham vehicle readiness checklists at the nearby Silverstone Grand Prix racetrack.

"You are Petty Officer Second Class Michael *McFinster*, newly assigned to the SUSLO?"

"*Mak-finn-skee*, and yes, assigned to SUSLO, Mr. Lantham."

"Please call me Terry." It was obvious that Terry bristled at the prospect of a generation gap, and wanted to establish friendly professional ties.

"Our security safeguards rely heavily on personal communication, and to build a good foundation, I need to know more about you. Are you married... children?"

What began as pro-forma chit chat soon transformed into a conversation delving into my professional background, international travel experiences, sports interests, commute practices from Cheltenham to the GCHQ, etc., etc. At times the conversation rambled. Terry was not taking notes, nor was there any kind of a recording device as far as I could tell. Tom Spivey would also toss me a question every so often, while he occasionally responded to an issue at the turnstiles. After about 20 minutes Terry asked for my badge, took out a clipboard with a form on it and wrote down my name and badge number, along with my new organizational code, room number and my supervisor's name and phone number.

"Petty Officer *McFinster* – you can please proceed to the Northwest side of the building, room 2557, on the second floor, where you will find Chief Petty Officer Bourgeois."

I thanked him while shaking hands, and headed to the Northwest wing to meet up with my new boss, Chief Bourgeois, a.k.a. "*Frenchie*".

Not only did Room 2557 have no cipher locks, there was daylight coming in through WINDOWS that could even be opened to let in the abundantly fresh Cotswolds' breeze. From the looks of it, the Brits had no use for the kinds of sterile, fluorescent-lighted windowless workspaces that I had become so used to. There were window sills and potted plants on the far side of the room, and teacups hanging from hooks over a kitchenette niche at the other end. Eight sturdy metallic work-desks were arranged in an open space that connected to a doorway leading to the Section Chief's and to CTRC Bourgeois' office, where he sat as the SUSLO Deputy to the Section Chief.

I made my way to a desk by the windows, occupied by a strikingly radiant young woman. She looked up from the papers on her desk and squinted at me, an unrecognizable "*Yank*" in a freshly-starched dress shirt, sporting a *horribly* inappropriate necktie (it was my UK Nimrod Squadron-51[119] organizational tie, one that I had picked-up as a souvenir when flying a Baltic mission series out of RAF Wyton[120], where Squadron-51 hosted our VQ-2 Detachment – *her squint was telling me that wearing an organizational tie from a unit that you don't belong to is a serious faux pas)*.

"Hello – I'm looking for Chief Petty Officer Bourgeois. Is he in?"

"And you are...?"

"Mike Makfinsky, a new report to Chief Bourgeois."

"Ah, yes. Well... Chief *Bour-ge-ois* (*boohr-ghgh-zhwaahzz*) is in a meeting. I'm Liz, the Senior VMF analyst." We shook. "Care for some Tea, do you?"

---

119 http://www.raf.mod.uk/rafwaddington/aboutus/51squadron.cfm. Retrieved on Jan 2, 2016. Squadron Motto: "Swift and Sure". The Nimrods were replaced by three Boeing RC-135W Rivet Joint aircraft in 2014. In January 2011 personnel from 51 Squadron began training at Offutt Air Force Base in the US for conversion to the RC-135. Crews will deploy on joint missions with the USAF 343rd Reconnaissance Squadron until the new aircraft are available. The first RC135W (ZZ664) was delivered ahead of schedule to the Royal Air Force on 12 November 2013, for final approval and testing by the Defense Support and Equipment team prior to its release to service from the UK MAA.

120 http://www.raf.mod.uk/rafbramptonwyton/history/index.cfm. Retrieved on Jan 2, 2017.

"Thanks, I'm usually caffeine-starved this time of the morning."

Liz, taller than I in her heels, moved with the grace of a Russian ballerina. She showed me over to the kitchenette, put water in a tea pot, and switched-on its electric heating element. Within less than a minute its contents were vigorously steaming. A pouch of Earl Grey went into each cup to be bathed in bubbling H2O.

Her Earl Grey pouch extracted-wrung-out-over-a-spoon-and-discarded, she folded rich cream into her cup. My guest cup vied for attention. Ditto – I wrung out the pouch and asked whether there was any lemon juice.

"LEMON? Why this is English Breakfast Tea... the thought of it!" "You *Yanks*!"

The tea was bitter... definitely needed a good measure of citrus to be palatable.

She again stared at the emblem on my necktie – a red Condor in flight over the number 51 in yellow, both emblazoned on a deep dark blue. I remembered the haberdasher who sold it to me out of his store in St. Ives (just outside of RAF Wyton Air Base), when I told him that I wanted it as a souvenir from my stay with Squadron 51... "*you Yanks*", he had remarked).

"Do you wear that because you're 51? You don't look quite that old."

I laughed, sensing that I was being measured up.

"Sister squadron – the emblem represents the Nimrod SIGINT squadron at RAF Wyton. I flew several series of Baltic missions out of Wyton. Ever been?"

"Can't say that I have, Mike, but I'm definitely quite fond of *Nimrods*[121]."

I was all ears.

"This is the Baltic Fleet Analytics Desk Office, more affectionately called the BA-FL-AN-DO. I've worked the BAFLANDO since starting as an Intern back in 1975, a year before graduating from University with a degree in Journalism."

I was impressed. "I've spent several thousand hours in reconnaissance aircraft, monitoring and reporting on Soviet VMF operations. Do you speak Russian?" I asked.

"Not one of my *forte's*. The Russian is translated in another department – BAFLANDO gets the final versions to then collate with information from other sources. What we call "Multi-INT" Processing and Reporting."

A stocky, pipe-smoking bespectacled gentleman in his mid-30's, sporting a salt-and-pepper full beard, busted into the room, leather portfolio tucked under his left arm.

"Petty Officer Makfinsky – I'm Chief Bourgeois. Welcome to the Team! Now that we've cleared the formalities, call me "Frenchie" – hell, everybody does, including my *Miss-US*! I already got your name from the Rota crew, Mak. They had nothing good to say about you."

We shook hands; he set down his portfolio and reached for a cup to brew some Nescafe instant coffee.

"Liz probably told you that we fuse COMINT with Multi-Source, and believe me, there's a method to the madness. It's kind of like what

---

[121] Nimrod – Old testament King, a mighty hunter, associated with ancient Mesopotamian hero Gilgamesh, the king who did not want to die. https://www.britannica.com/biography/Nimrod. Retrieved on Jan 2, 2017.

we do at FOSIF Rota, only here at GCHQ we take in a lot more information. We have dozens of analytical reports dangling on the vine, waiting for just the right kind of HUMINT to come in as collateral. Can't put out anything that's half-baked... remember **that**: my word to the wise, right Liz? By the way, Mak – Liz is our Editor in Chief. We Yanks are piss-poor when it comes to grammar, syntax, and style in general. Liz has an OVER-abundance of all of these components."

"Like you often say, Frenchie – good reports, like your Mom's apple pies, need ALL of the ingredients for perfection. In our case it's SIGINT (*Signals Intel*), HUMINT (*Human Intel*), IMINT (*Imagery Intel*), GEOINT (*GeoSpatial Intel*), MASINT (*Measurement and Signature Intel*), TELINT (*Telemetry Intel*), EDINT (?) and when available, a touch of WIZINT (?)."

"ED-INT... WIZ-INT?" I knew there was a punch-line lurking.

"She's referring to Editorial Intelligence and Wizardry Intelligence – sources that only Liz can capitalize on."

It was a good start – I liked the team, and really enjoyed the Hail party that the Officer in Charge, Lieutenant Commander Rickenbacher and his lovely wife Jenelle held for Victoria and me in their secluded Cotswolds country cottage – an elegantly renovated 17th century farmhouse. Victoria saw it as a throw-back to Colonial Britain – something to be relegated to the past... she was (and *IS*) all about minimalism and functionality. That said, we would have moved into a like home in a heartbeat had we been able to afford it. The one thing that didn't work at all for Victoria and me in Cheltenham, and which became a big obstacle to acclimating, was the lack of permanent housing for our family of five.

THE Cotswold Grange Hotel

The Cotswold Grange Hotel (CGH) had no double-adjoining rooms, so the only room available for a family of our size was one in which a walk-in closet had been converted into a breakfast nook with built-in benches. The Hotel management removed the breakfast table, put two futon mattresses on the benches, and *PRESTO*: a bed for Ivan, who was 5, and a bed for Paul, who was 3. Dan, our youngest at the age of 2, was to sleep in a crib alongside Mom and Dad's queen-size bed. We were living the dream:

The Little Hotel Room on the Cotswolds.

What had been lacking was truth in advertising about my orders. While my family was in fact "Command Sponsored", which meant that the Navy was on the hook to guarantee suitable housing, what had not been factored-in was that one of the SUSLO staff members had decided to EXTEND his (and consequently his family's) tour of duty at the GCHQ, and that it was HIS HOUSE that had been earmarked for my family to move into within a month of our arrival.

Two months went by, and NO HOUSING to be had.

After a week into the THIRD month, STILL NO HOUSING TO BE HAD.

While I had work to keep me busy, and had been invited to join the Cheltenham College Squash and Tennis Club (thanks to a special arrangement that GCHQ had with the Club, I quickly incorporated into GCHQ's tennis team), Victoria, (as always - tough as nails), was really getting jaded with the Hotel situation. The boys were already going BONKERS because, truth be told, the CGH was NOT a child-friendly establishment. Ivan, Paul, and Dan were the only children at the hotel, and because breakfast, lunch, and dinner were included in our full board rate, we had a table permanently reserved for us toward the back of the dining area, as far away as possible from the adjoining Bar. In 1980's Britain, Bars and Pubs were *allergic* to children. Victoria and I really resented not being able to plop down at good rest stops on our all-too-frequent countryside drives with the boys, since the best food was always to be found at quaint, culturally attractive roadside pubs, rather than at the occasional KFCs or McDonalds (*Wimpy's was an exception – the boys would cause such a ruckus over a Wimpy's sighting that it almost always became a mandatory stop*).

He Wanted to FLY

As our stay at the CGH progressed well into MONTH THREE, I came home to a mind-boggling surprise one evening, after parking the car outside the hotel's front yard.

"Mr. Makfinsky – there's a police constable wishing to talk to you." Said Sarah, the owner-manager-chef and all around superb hospitality-industry guru.

"Yes – Mr. Makfinsky, one of my colleagues is getting your wife Victoria's story, but I want to get some facts from you directly." The constable was low-key, matter-of-factly, as only British cops can be.

"And what does this concern?" All I could think of was that my car was not properly registered, although I had followed GCHQ's and SUSLO's guidelines for registering a left-side-driver "import".

"Fortunately there have been no injuries, and your son Daniel is perfectly alright. At approximately 4 pm today one of the hotel guests was having a drink in the cellar Pub, and he heard a loud thump coming from the outdoors patio. When he ran up the steps out of the cellar and onto the garden patio to investigate, he saw your son Daniel sitting on the edge of the patio's Plexiglas awning. The poor little boy was wailing uncontrollably."

"I need to see my son and my wife... immediately!" was all I could say, knowing that this had probably happened while I was "homeward-bound" from the GCHQ.

"They are perfectly well, I assure you. It appears that your boy found a way to open one of the windows on the winding staircase, which is not surprising, since this hotel is not child-proofed, and that he managed to wiggle his way out of the window to fall from a point about mid-way between the first and second floors, directly onto the Plexiglas awning over the garden-pub. What we have not been able to establish so far, is whether someone, a brother perhaps, might have been playing with him at the time, instigating, if not causing, the accident."

By now I could tell that this was a BIG DEAL at the CGH, which had up until now fit the Oxford Dictionary's definition of a sleepy hollow. Most of the residents had gathered in the garden pub, and were discussing the "main event", speculating as to whether the Mother had neglected the child, or perhaps even had something to do with this mishap... an attempt to teach the little rascal a *LESSON – eh?*. Yes – the family has been cooped up in that room for over two months now, and this is what happens...

"Mr. Makfinsky – my apologies, sir." - Richard was Sarah's husband, co-owner, co-manager, and possibly Head-Chef- "Constable – can I have a word with Mr. Makfinsky?"

"In due course, sir – I must remind you that one of the issues here is whether the Hotel is safe for children."

"Now, Mr. Makfinsky – one of the questions that needs to be answered is why are your boys NOT in child-care, nursery, or kindergarten. Well?"

I gave a short explanation, what with the temporary lodging penance we had been serving ever since I had reported to my assignment at the GCHQ.

"Aha! You're at the GCHQ… I know for a fact that American employees at the GCHQ have assigned housing near the Benhall campus. It will be ideal for your boys… I suggest that you move there **immediately**."

So now the short explanation did not suffice. I gave the Constable both barrels-full, and as I rambled, Victoria finally came into the picture, along with Ivan, Paul and Danny – whose face was all red, making it obvious to me that he had no tears left in the ink sac. I picked Danny up, and hugged him in a reassuring way: No… you're not in trouble, Mr. Danny…

Victoria was *flapped*, and when *flapped*, her preferred mode of expression is in Castillian Spanish, which I'll translate directly.

"All four of us came downstairs to go for a stroll, and as I was talking to Sarah about which new area we might explore, Danny must have stayed back. I got sidetracked putting Paul's coat on, and when I turned around, Danny had disappeared. The boys and I went looking for him in the dining area, then back to the lobby where I asked Sarah to mind them so I could check the room. As I ran back up the stairs, I noticed the open window, and could hear Danny crying, so I ran downstairs to get him. One of the hotel guests, that kind fellow over there, was holding him by the hand. Danny's Ok – a bump on the back of his head, and a scratch on the palm of his right hand…" She looked tired, and infuriated by the mishap.

I suggested to the Constable that we take a good look at the window in question. He agreed. Richard joined us as we climbed the winding translucent staircase, its walls made of corrugated Corning glass tiles – it was impossible to discern anything through these tiles, other than the general outlines of the trees-beyond.

There it was, at a point about half-way between the first and second stories: a steel-framed 1′ wide x 2.5′ high window made of corrugated glass tile with a very simple latch. The window swiveled on an axis that was about three quarters of the way into the window-pane's width – an opening wide enough for window-cleaning. Only janitors and small children would take notice of that latch.

Could a two-year old child have opened the latch and "jumped" out of a 9-inch wide, two and a half foot high opening? Maybe not, depending on how tightly secured it was. Assuming the latch had been left unfastened, or that the window was ajar, however, it *was* possible for a toddler to push it fully open, and then to fall through it from the momentum used to shove it open. The window offered no resistance whatsoever once the latch was unfastened. Since Danny couldn't articulate his side of the story, forensics would have to include fingerprinting to see whether Dan had opened the latch... good luck fingerprinting a two-year old!

We looked out the open window, and about ten feet down below was the yellow-tinted Plexiglas awning, covering the garden-pub. It was not a heavy-duty awning, but sturdy enough to keep a 25-to-30-lb. child from crashing-on through. Since it was plastic, and was anchored to the patio floor with an array of plastic-sheathed aluminum tubes, it had some give, in this case enough give to cushion the fall. Fortunately there was no trampoline effect upon impact, and Danny was thus spared from being launched back up and over its edge.

"This proves" –said the Constable- "that this staircase is unsafe for children, and this window, and any others like it, needs to be welded-shut if these boys are to continue to reside here."

"It's lucky that your boy didn't crash into it head first... it might have given way!" Richard was scratching his head, as was I, grimly pondering what might have been.

Ivan and Paul blurted out what, in their minds, was at the heart of this catastrophe averted:

"Daddy – Danny wanted to FLY!"

But just what was going on in Danny's head, I wondered. I looked Danny in the eye as he held back non-existent tears, and growled internally at his flashback. If only he could talk...

"DARN IT MOM! - Here we go again on another one of those never-ending walks to a playground with monkey bars and swings that I can't use (ok, I must admit that the teeter-totters are ***FUN*** when **you** push, Mom, and **you** do all the work for me... heck, as much as I hate to admit it, I'm still a *littl*e critter!). Sure, Ivan and Paul get to mosey alongside you, but when it comes to me, I either get stuffed into a stroller, or yanked by the arm when I want to stop and check out the bugs. No, not today, **Mom**! I think I'll hang around the Hotel and *chill*. Maybe I can give you guys the slip when you're not looking. Yeah... keep on talking to Sarah... here I go, up the staircase *sly as a fox I go*.... Stay low, laddie... you can do it. HAHA! They're looking for me in all the wrong places. I'm pretty sneaky. YES I AM! This is **sooo good** that I had better find a nice vantage point to take it all in. Back down the staircase to get a better view... that's it, bump my rump one... bump my rump two... bump my rump three... bump my rump four.... Right here! They're wandering through the dining area now, searching for me... all in va-va-va-vain! Sarah is nowhere to be seen. GREAT SPOT, this is. Not only can I see the lobby and the hallway that leads into the dining area, I also have this *COOL* back-lighting from the corrugated glass tile behind me. ? ; -) ? How about I strike a *pose?*... Yeah, one of those *Fred-Astaire-dashing-leading-man-I'm-the-Numero-UNO-gift-to-the-WORLD* kind of pose! They will find me leaning, one hand against the glass wall, my toddler feet crossed, one heel pointing right at 'em, and to top it off, the *piece de resistance*... my most accomplished full frontal-dentured Cheshire-cat cheese-eating grin, switched to *MAX* to bathe 'em in glamour. Yep. That's what I'll do. When Mom finds me looking like a Greek statue, having outwitted ‘em ALL, it will make her day to look up and see me here, **VICTORIOUS!** No doubt that if Dad were here, he would be getting all of this *Oscar-wiener-quality* performance on his Super-8 camera! Ok, so extend the right arm like so, and the right hand goes here… **WHOOOOOPPPPSSS!**

***NOT GOOD!!!*... TALK TO ME GOOSE!**

Jeez, this feeling of weightlessness is kind of like when Dad tosses me up in the air…

WHEEEEEEeeeee

So this is what Einstein means about the time and space continuum…

Plexiglas, HERE I COME…

KA-THUNK!!!

The episode caused some consternation to the owners of the CGH, and given our housing uncertainty, Victoria soon resolved to fly back to her parents’ home in Madrid. She really needed a hassle-free Summer vacation, so her July departure was to cheer up the boys who would soon be living large once again in their grandparents summer house in Collado Mediano[122], on the slopes of the Sierra de Guadarrama, Northwest of Madrid.

Finally: Officer Candidate School, BUPERS FY-80 Orders Ser. Nr. 0013289

At right about this time I got news that I had been selected for Officer Candidate School (OCS)[123], a four month “knife and fork” program

---

[122] Collado Mediano – offers quaint horseback riding through mountain trails alongside Roman and Moorish ruins. https://www.yumping.com/ofertas/rutas-a-caballo/madrid/iniciacion-a-la-equitacion-ruta-a-caballo--o10462

at Newport, Rhode Island. Silver Fox was hammering out orders for me to report to the next class, one convening in late July. The re-assignment process was complicated because I was not eligible for a Permanent Change of Station (i.e. *PCS Orders*) unless I was assigned to the Naval Station at Newport Rhode Island for at least seven months. As luck would have it, CTICS Yeargain kludged together a set of Franken-orders for my transition from Enlisted man to Commissioned Officer, orders that included some time in limbo with NAVSTA Newport's First Division, followed by four months in OCS, followed by several Officer Basic Schools, one of them being Communications Officer Ashore. A question mark remained over where I would go upon graduation from the Communications Officer Ashore course, since by then the Silver Fox would have no control over my fate. Depending on my officer designator, an Officer Detailer would take control, and issue me my next set of orders. For all I knew, I could start my officer career as a *"butter-bars"* Ensign at a Naval Communications Station in Point Barrow, Alaska.

This new career development came on the heels of an inability by the Navy to make good on a Navy Enlisted Scientific Education Program (NESEP) 2-year scholarship I had been selected for back in January 1980. The NESEP office had slated me for an Electrical Engineering degree at the University of Washington to commission as a Navy Unrestricted Line Officer (*Unrestricted Line Officers can take command of a United States Ship, whereas Restricted Line Officers cannot*[124]). As luck would have it,

---

[123] http://www.ocs.navy.mil/ocs.html. Retrieved on Jan. 2, 2017. The U.S. Navy's Officer Candidate School (OCS) is located at Naval Station Newport in Rhode Island. The 12 week OCS course is designed to give you a working knowledge of the Navy (afloat and ashore), to prepare you to assume the responsibilities of a Naval Officer, and to begin developing you to your fullest potential. OCS is extremely demanding; morally, mentally, and physically. Your personal Honor, Courage, and Commitment will be tested at OCS and you will be challenged to live up to the highest standards of these core values.

[124] There was one notable exception to this rule, when in the 1980's a "Restricted Line" Cryptologic Officer was put in charge of the USS Constellation Battle Sloop as she awaited administrative transfer from the U.S. Navy to the City

the NESEP program was defunded by Congress within the very same month that I was notified. On that note, Lieutenant Commander Paul Christensen ordered that my NESEP application be revamped to conform to an application for OCS, which he summarily had me sign for expedited delivery to the Bureau of Personnel.

To get Victoria and the boys started on their well-deserved summer 1980 vacation, we caught a "Space Available" flight on a Military Airlift Command (MAC) aircraft from Mildenhall Air Base in the United Kingdom, to Torrejon Air Base near Madrid, Spain. We had never seen or even heard of the C-47 that was advertised on the flight notice, and I suspected that we were in for a surprise. Sure enough, when we were bused out to the tarmac, there stood the C-47, a military version of the very same DC-3 that appeared in Humphrey Bogart's "Casablanca" movie (remember the final "Here's looking at you, Kid" scene[125], when Bogart bids farewell to Ingrid Bergman who is to board a DC-3 to high-tail it out of Morocco?)

---

of Baltimore in order to become a permanent museum ship in Baltimore's Inner Harbor. Her history includes noteworthy events, e.g.: Re-entering active service in June 1859 as flagship of the African Squadron, Constellation took station off the mouth of the Congo River on 21 November 1859, she captured the brig Delicia during the mid-watch on 21 December 1859 "without colors or papers to show her nationality... completely fitted in all respects for the immediate embarcation [sic] of slaves..." On 26 September 1860, after her entire crew had turned-to to "trim the vessel for the chase" (even wetting the sails "so they would push the sloop along"), Constellation captured the "fast little bark" Cora (which showed no flag and carried 705 slaves), nearly running down the slaver in the darkness. When captured, the slavers were impounded and sold at auction, their captains required to post bond and await trial, while their crews were landed at the nearest port and released. The newly freed slaves were taken to Monrovia, Liberia. The U.S. government paid a bounty of $25 for each freed slave freed, and "prize money" for each impounded ship to be divided among the crew proportionally according to rank.

[125] https://www.youtube.com/watch?v=rEWaqUVac3M Retrieved on Jan 2, 2017. Final scene of "Casablanca".

We climbed aboard the C-47 *Skytrain*[126] and were impressed from the get-go by the friendly aircrew, and the impeccably maintained aircraft of polished stainless steel fuselage. As we were the only passengers (why more GI's stationed in gloomy England did not want to "hop" over to Madrid, Spain in July is incomprehensible to me), after take-off and when we had gained altitude, our three boys, Ivan - 5, Paul – 3 and Dan – 2, romped around as if they were in an "air-park", the many empty passenger seats providing good hiding places for a great game of tag, and the many windows offering terrific views of the clouds, coastlines and terrain contours below. As the Skytrain cruised South at an altitude of 18,000 feet, Mom and Dad couldn't have asked for a better flight, and we even joined the boys in their fun and games. Our route South took us over Brest, France on the Western tip of Brittany (*Bretagne*), then again into the Atlantic, and finally over Spain to reach Torrejon Air Base (near Madrid, Spain's capital) via an itinerary that tracked near the ancient cities of Santander, Burgos, Aranda de Duero, and finally Alcala de Henares as we made an approach from the Northeast to land on the

---

[126] http://www.flyingmag.com/dc-3-an-airplane-for-ages. Retrieved on Jan 2, 2017. Eighty years after its first flight in December 1935, the DC-3 remains relevant and revered because it fulfilled so many roles, carrying millions of passengers aloft for the first time, and playing a key factor in turning the tide of World War II for the Allied forces. Each mission required a tweak to the airplane's basic capabilities: a sturdy airframe, built to withstand both the rigors of commercial transport and the abuse in far-flung climes, and a forgiving aerodynamic nature, temperate enough to suffer the ham-fisted efforts of greenhorn pilots and reward the fine-tuned handling from experts willing to coax more from its burly engines and, at times, recalcitrant systems.

Torrejon Air Base[127] Runway 222. I looked out the port window onto the Torrejon Air Base Golf Course, one that I had played as a Junior and Senior in High School. The Golf Club snack-bar was known for serving the best burgers and fries on base (possibly the *BEST in Spain* in the 60's and 70's), to be chased down with San Miguel beer in longneck bottles. Later in my career, as a Naval Officer, I would get to know the Philippino version of Spain's San Miguel beer, affectionately called "San-Magu" by sailors and scallywags alike.

The family reunion was memorable, and the retreat in the Sierra de Guadarrama countryside proved to be just what Victoria and the boys needed after being hemmed-in at the CGH. Victoria's parents would take the boys out on walks through the Sierra's winding dirt roads, where they stared-down cows, and ran from bulls to their hearts' content. To acclimate them to their new "cattle ranch" environs, *Abuelo* (grandpa) outfitted them with toy six-shooters, holsters, and cowboy hats & boots. Danny became so fond of the boots that, unlike his brother "Tex" Ivan, our only true cowboy by birth, he would not part with them, even at bed-time. "I'll die with my boots on!" was Danny's Guadarrama mantra.

Victoria, Ivan, Paul, and Danny were back in the groove, recharging spent energies and living large – La Buena Vida Navy once again.

NAVSTA Newport RI

I caught a flight from Madrid to Boston, rented a car to haul my sea-bag to Newport, and reported to the Naval Station Newport Rhode Island Quarterdeck.

"CTI1 Makfinsky reporting for duty." I handed over my set of MILPERS orders (labeled "Original Copy") to get an official arrival time-

---

[127]http://www.ejercitodelaire.mde.es/ea/pag?idDoc=BD09B2FC0FBC4052C1257 4480039C7ED#80A16BC8A3F5BB62C125745E0030FEA8. Retrieved on Jan. 2, 2017. Torrejón de Ardoz (Spanish). Homebase to Fighter Wing 12 (F-18).

hack from the Officer of the Day (OOD). Without that time-hack, my travel was not considered completed, risking administrative dereliction of duty.

"Welcome aboard, First Class Mah-finsky – I have you reporting to First Division tomorrow. You'll stay at the BEQ (Bachelor Enlisted Quarters) until your OCS Class convenes." The OOD, Senior Chief Boatswain's Mate (*Bo'sun*) Harvey Bussel (pron. *Bew-Sell*) asked whether I needed the duty driver.

"Not now, Senior Chief – I can get to the BEQ, but will have to return the rental at the airport. Could use a duty driver to help with that."

"Then I'll have the duty driver meet you at the BEQ in an hour. You'll report to Master Chief Standage, First Division's LCPO at 06:30 tomorrow, in building 369, next to the flagpole near the pier."

The car returned, and back at the BEQ, it was time to meet some of my NAVSTA Newport shipmates before assuming the first day watch at 06:30 the following morning. And where better to meet the locals than at the pool table in the BEQ lounge. It became obvious to me who the lounge lizards were: a tall, lanky guy twirling a pool-stick while his opponent considered how to get out from behind with a low-percentage shot, and a short stocky bruiser plumped down in the EZ-boy chair near the door, pretending to watch TV.

Barney, the tall pool-shark, was a Bosun onboard the USS Connole FF-1056 / "*Exempla Suorum Durant*" The Example of Our Ancestors Endures /, a Fast Frigate that had taken-up home port at Newport just the year before. Trevor, whose short-sleeves were overstuffed from a thick pair of guns, was a SEAL by trade, and a Machinist Mate by training, serving a tour at Explosive Ordnance Disposal Mobile Unit TWELVE Detachment.

"Anyone up for a $5 8-ball challenge?" I looked toward the TV, then the pool-table.

"Table's taken, dude – gotta wait your turn for a thrashing." Barney was used to talking trash. "You're new?"

"Yup – just got off the turnip truck, in a holding pattern for OCS." I answered, staring him down as a prelude to turning his luck at the table. "I'm Mike Makfinsky – CTI1, Russian linguist. Mak for short."

"Oh, yeah – one of them OCS prima donnas, right? I'm Barney Cairns, Bosun Second, USS Connole, and my buddy yonder is Trevor Santos. He thinks he's a SEAL... BARK, Trevor!" Trevor didn't budge; he just raised his right hand, finger wave back at Barney.

Barney carefully chiseled his next four shots, then slammed the eight-ball into a side pocket. "Next." It was my turn to pony-up some greenback. I put a $5 on the table and proceeded to rack.

"Lag for the break?" I suggested.

"Negatory, Kimmo Savvy – you challenge, you get sloppy seconds." Barney held onto any and all of the advantages he could get. "Matter of fact – how's this for a break?" He piddled the cue ball toward the rack. It barely struck the lead ball, and only three balls trickled from the "rack".

"You CT's break like that?"

"Nice, Barney – Negatory to ya. F-Y-I, that's a foul[128]. Now you get to rack, and I'll break. How d'ya like them apples?"

---

[128] http://www.wpa-pool.com/web/index.asp?id=114&pagetype=rules. Retrieved Jan 3, 2017. If the breaker pockets no object ball, at least four object balls must be driven to one or more rails, or the shot results in an illegal break,

"WRONG – no foul. You gotta play 'em like they lie."

The situation was about to get a'-cracklin' so the SEAL, smelling adrenalin woke from his slumber.

'Barney, you sorry sister – he's RIGHT! That's no break! Gotta have at least FOUR balls pop and hit rails. What a piece of work you are, Dude!"

Barney slammed his pool stick on the table. The rap was audible throughout the entire lounge and beyond. "Goddamm it, Trevor – keep your snout out of this! There's no racking to be done. The table's open, and Mak, you play it as it is!"

This was no longer about 8 ball, it was all about PSYOPS (*psychological operations*), and the tide had turned my way.

"Ok, Barney – here's how it goes... you just fouled again. Cue stick can't touch the table... in 8-ball, you "double-foul" and it's over. You owe me FIVE, amigo!"

Before Barney could unleash his rage, Trevor stood up. "It's MILLER TIME, LADIES!" He slung a weathered biker jacket over his shoulder, and made toward the door. Barney threw a $5 onto the pool table, clipped the cue stick into the wall mount, and followed Trevor. Of course, I pocketed the $5 and tagged along, knowing that for Barney this was not a good start (there was no doubt in my military mind that I would have lost, but for the "rules") although the stubbed toe would soon be forgotten after a few brewskis at the Club. As Nikkie New Guy, the first round was to be on me. We loaded into Trevor's metallic blue Camaro 350-SS.

---

and the incoming player has the option of (1) accepting the table in position, or (2) re-racking and breaking, or (3) re-racking and allowing the offending player to break again. During Play: If the shooter commits a foul, play passes to his opponent.

It was not an impressive E-Club, but it was by the piers with a good view, and a deep New England ocean breeze prevailed to flare the nostrils. It was cavernous, dark, offering many tables, with music blaring, alternating between Rock and Country. This was a week night, but still, some of the local girls were carousing, and Trevor's track-while-scan instinct was on.

"Check it out – that table over there's got some stone cold vixens." Trevor sat himself down in the immediate neighborhood of the alluring perfume. "Hi – I'm Trevor, and these two Neanderthals have no names."

"Well, Trevor, ain't you just the cutest spark to fire my plug – I'm Cindy, but call me *Candy*, and these three girls are my cousins. They're *ffvvisiting*. I live *just* down the road, *you know*?"

Trevor had struck pay-dirt. Cindy, rather *Candy*, was on the prowl, and ready to shake her "cousins" for some deep tissue massage, preferably coming from a studly SEAL. Pitchers of Narragansett beer were delivered to our table, while frozen vodka shooters were poured into 6-oz. glasses at their table. After several rounds, the two tables came together.

"Candy – you're definitely not the "sugar-free" variety. I gotta tell you, *you syrupy Schwöbli*[129], that I'm not NOW, nor have I *EVER* been on a diet."

Not great on the wit, but with plenty surplus on the brawn side of the equation, Trevor was on a roll. It was time to play his SEAL trump card. In these circles, that always aroused some wet pouting between the legs.

---

[129] Schwöbli - ORIGIN: Switzerland. DEFINITION: A bun that highly resembles a bum. Trevor had trained with the Swiss Special Forces Army Reconnaissance Detachment 10.

"A ***Syrupy SSHHHWAYBLEEeee?*** How cute, Trevor. And just what does that mean, something *sexxyyie*? Gotta say that everything coming from you sounds sexy, you ***over-under*** *Beretta 686 Silver Pigeon*, you!"

"Ok, guys – get a room." - the brunette cousin winked. She was trying to focus on Barney.

"Rather than explain, let me show you how a Schwöbli is eaten, one voracious bite at a time. There's a reason why we SEALS are called *snake eaters*!" And with that, Trevor chomped a generous bite out of his beer glass, and proceeded to grind away at the shard of glass slowly, deliberately, making silicon powder with his molars.

"Holy Jesus, Trevor – you're at it again!" Barney shook his head, proof that he had seen his SEAL friend pull this prank one time too many.

Candy leaned over to give her new-found hunk a well-deserved lip lock in reward for this **extreme** bravado, and *bada-bing*, the still unprocessed shard cut Trevor's lower lip. He grabbed her glass of vodka, took a swig, and in the process turned the drink deep red. The cut must have been a good one. Still, nothing for a *Snake Eater*.

It was time for me to move on. I was only four hours away from my wake-up call, and I could tell that this party was headed to some undisclosed location "just down the road". I expected NOT to run into either Trevor or Barney the next day, and hoofed it back to the barracks.

Not much to say about my short stint with First Division – as a First Class PO, Master Chief Standage used me to lead some base security patrols, and even sent me out into Newport a couple times to settle bar disputes between NAVSTA sailors and locals. They were easily resolved, with the sailor paying for damages, apologizing, and riding back in the wagon with my Patrol to report to the Master-at-Arms (MAA). This was a far cry from what Navy MAAs dealt with in foreign ports, where the occasional bottle-smashing brawls and knifings resulted in police blotter entries. This was one last grease-mark on my enlisted slate: Leading Petty Officer for Shore Patrol duties in Narragansett Bay, RI.

My OCS Class convened on a Sunday. Officer Candidates were required to report to the Quarterdeck No Later Than (NLT) 19:00 (7 pm), allowing for one last dinner with friends and family. While technically I remained a First Class Petty Officer, CTI1 Russian Linguist for the duration of OCS, in a more practical sense, I was just another one of the hundreds of Officer Candidates, ready to be indoctrinated into the Commissioned Officer echelons of the U.S. Navy.

My stint as a non-commissioned officer would soon be over, if only I could get though the rigors of Officer Candidate School. Could OCS be any more demanding than was that 1974 Boot Camp?

Appendix – CIA Description of the EP-3E Aircrew

In the EP-3E, the crew consists of 28 members while the EA-3B carries a 7 man crew. To identify the unique talents of its enlisted and officer corps, the VQ squadron utilizes the following terms: Electronic War-fare Mission Commander (MC); Electronic Warfare Aircraft Commander (EWAC); Electronic Warfare E valuator (EVAL); Electronic Warfare Navigator (EWAN) and the Electronic Warfare Aircrew man (EWOP). The following specifically define the duties, responsibilities and levels of proficiency of each position.

DESCRIPTION OF POSITIONS

Mission Commander (MC). The designation of Mission Commander is reserved for select individuals, who by virtue of their extensive knowledge of the principles of electronic warfare, squadron aircraft operations and crew coordination have been designated by the Commanding Officer as the person ultimately responsible for the conduct of the ESM mission. This responsibility makes it imperative that the Mission Commander maintain full awareness of every aspect of the intelligence collection mission, so that all phases are carried out in a comprehensive and coordinated manner. The Mission Commander briefs the crew on specific responsibilities and special taskings in light of current political situations. Once airborne, his complete understanding of complex aircraft systems and the signals environment, as well as his superior ability to direct the crew effort, enable him to complete the mission and to optimize the intelligence gain. His decisions are based on available information and he knows a wrong decision may suddenly place the naval forces in jeopardy. On the ground he supervises all first echelon analysis of the mission product and has sole responsibility for its accuracy and timeliness. This final step is critical since his interpretation of situations may have national and international ramifications.

Electronic Warfare Aircraft Commander, EWAC. The Electronic Warfare Aircraft Commander (EWAC) is a pilot with a high degree of maturity, experience and skill. Furthermore, he must be able to perform professionally under stress. Routinely, his authority greatly outweighs that of his contemporaries in other fields and he bears responsibilities not usually required of other type aircraft commanders during their entire naval career. The EWAC must be constantly aware of the political sensitivities in the area in which he is flying and be prepared to make instantaneous decisions concerning situations which, without the proper reaction, could easily develop into an international incident. He is also capable and responsible for relaying vital intelligence information directly to Fleet Commanders, CNO, NSA or JCS as the situation warrants. In the absence of a designated Mission Commander, the EWAC is responsible for coordinating Electronic Warfare (EW) tactics with the Electronic Warfare Tactical Evaluator and the Electronic Warfare Navigator.

Electronic Warfare, Tactical Evaluator, (EVAL). The Electronic Warfare Tactical Evaluator (EVAL) manages the planning, collection and reporting requirements demanded by each mission. The EVAL must collect signals intelligence (SIGINT) to determine the evolving tactical scenario. This collected information is compiled into a SIGINT report which reaches the highest echelons in the national reconnaissance efforts. The political sensitivities inherent in the various areas of operations require the evaluator to be completely knowledgeable in areas of U.S. and foreign strategy, tactics and national objectives. On numerous occasions, a squadron aircraft has provided the sole input of national security information which reached the national command authorities rapidly. It is evident that in addition to a high level of expertise and leadership, the EVAL must possess a high degree of maturity and judgement. The EVAL carries the ultimate responsibility for the analysis of intelligence during airborne reconnaissance missions in the EP-3E and EA-3B. To be designated a Senior Evaluator (SEVAL), a Naval Flight Officer must demonstrate his ability to perform a SIGINT collect mission in any tactical environment. The SEVAL is considered a tactical SIGINT expert and as such is routinely called upon to provide high level briefings and assist in the development of Electronic Warfare tactical doctrines.

Electronic Warfare, Navigator (EWAN). The Electronic Warfare Navigator (EWAN) is an integral member of the mission crew. His qualification as a navigator requires a complete understanding of several navigational systems. Navigation must be constantly checked and maintained within a one mile accuracy. This is necessary to ensure the preciseness of the mission product and to avoid international incidents when flying near hostile countries. The EWAN must complete all NATOPS requirements for the type aircraft flown. In the EA-3B, the EWAN is instrument qualified, completely familiar with all aircraft systems and must function as a co-pilot to enable the EA-3B to be categorized as a dual piloted aircraft. He additionally handles all radio communications from take-off to landing. Additional responsibilities include communication to and from the aircraft as well as recognition and photography of aircraft and ships. His knowledge of the mission is necessary for him to coordinate with the Aircraft Commander and SEVAL for mission accomplishment. EWANs must be aware of flight rules governing these highly sensitive missions. Missions in close proximity to non-friendly countries are conducted with complete reliance on the EWAN's knowledge of the rules for that area. Mission reports and the time constraints governing them are also part of his required knowledge. Finally, the EWAN must possess a high of experience in both Naval and Electronic Warfare.

Standard Aviation Electronic Warfare Aircrewman. The backbone of VQ-2's Electronic Warfare Crew is made up of the highly professional enlisted Naval Aircrew men. There are over 150 Aircrew men of various ratings and experience levels who directly determine the success or failure of the EP-3E and EA-3B missions. The flight engineers on the EP-3E and the crew chiefs on the EA-3B are drawn from the AD, AM and AE ratings. They are responsible for overall airworthiness of the airframe from preflight through completion of post-flight. Without their presence on the crew, the aircraft would never leave the deck. The lives and safety of the entire crew depend on their technical knowledge and professional readiness. The radioman's position is usually manned by an AT who must be fully knowledgeable of the EP-3E's Communication/Navigation systems. The Airborne Electronic supervisor, or "Tech", is a senior AT who is responsible for making sure that all the sophisticated Electronic Warfare equipment is in optimum operating condition. He is also responsible for crew coordination, on the deck as well as in the air, and therefore must be a mature and professional military leader. The lab operator is an Airborne Electronic Warfare Analyst. His complex task requires expert knowledge of ELINT as well as an in-depth working knowledge of the complex analysis and recording systems on the EP-3E. The Evaluator relies on the CTT for a successful mission. Most of our Naval Aircrewmen are Electronic Warfare Operators (EWOP) and make up the front line of ELINT collection. These highly trained AT's and AE's master the operation of sensitive receiving equipment as well as develop an in-depth knowledge of ELINT, a requirement not usually associated with this rating.

III. It's About What You Take With You

So what does all this mean? How does it translate, in terms of "the meaning of Life"? I recall how my good High School friend Steve Rodriguez was flabbergasted when he heard that I had "signed-up" for a hitch with the U.S. Navy in early 1974[130].

"Man, Dude – you've got to be kidding... I just can't picture you in such a large organization. And to stay locked-in for FOUR YEARS???" Steve slurred into the phone. I could hear Janis Joplin music blaring in the background –"Down on Me" was the tune[131], and I pictured Steve with one of his many transient girlfriends, perhaps toking on a bong, the two, or three, enjoying the balmy Madrid springtime breeze filtering in from his 10th-floor downtown Madrid balcony into his lavishly cushioned Tangiers-style living room. Steve was one of those rare *MHS-Knights* exceptions who had gone ex-pat after our graduation in 1971. He was capitalizing on his fluent Spanish as a free-lance journalist, bringing USA ground-truth to Spain's newspapers. In addition to journalism, he spent his time casually cultivating other pastimes, such as collecting the finer scents of beautiful adventure-seeking women touring Madrid's museums, restaurants, and concert halls, for whom he gladly served as a tour guide.

---

130 http://www.ibiblio.org/anrs/docs/V/1101%20US%20Naval%20Strategy%20in%20the%201970s.pdf. Retrieved Jan 10, 2017. U.S. Naval Strategy in the 1970's – a collection of Naval War College Newport papers, John B. Hattendorf, D.Phil., Editor. Provides background on U.S. Navy strategic posture in 1974.

131 The third and final stanza of Joplin's version of "Down on Me" ends with a positive message: "Believe in your brother, have faith in man, Help each other, honey, if you can Because it looks like everybody in this whole round world is down on me." https://www.youtube.com/watch?v=pVSOFI171NQ. Retrieved Jan. 10, 2017.

"It's time to move on, man. The world here in Madrid is way too small, and there are a lot of experiences waiting out there for us free spirits, Stevie."

"Yeah, but the *NAVY*? You've seen what military life is all about, dude, hammerin' away day-in and day-out for some war-mongerin' hawk. Plus, 'Nam is still cookin', man – you could get thrown into that hell hole if the peace accord back-fires. You know that the U.S. Senate hasn't even ratified it? Some say it's because Nixon wants to keep the door open in case he decides to put ground troops back into the fray. So, brother, nothing has been resolved yet, and the casualties keep climbing!"

I explained to Steve that I was signing up as a Sonar Technician, so chances were that I would be hunting down Soviet combatants in the Atlantic or Pacific, depending on where I got stationed.

"Ok, in that case, make it short, brother, 'cause while you're out there boring holes in the ocean, this World will keep on turnin', and you'll want to get back on it to catch up with the rest of us."

Steve definitely lived in his bubble, and I realized that now I myself was about to jump into a new bubble as a combat sailor in the United States Navy.

"No worries, Stevie – I'll do my share to keep the world safe for democracy so that you and yours can continue to live long and prosper." That was my sophomoric attempt at prophetic words, drawing from Star Trek's Spock, and I had a hunch that Steve was only a few hours, if not minutes away from yet another attempt at thriving in the ruckus of Madrid *la Nuit*.

Steve was indeed living the life of Riley, one that we had all aspired to after our HS graduation (as a matter of fact, coincidentally, in Spanish culture the mirror image of Riley is Rodriguez, so that "to live the life of Riley" in Spanish translates to *"estar de Rodriguez"*).

There are many reasons for becoming part of a much larger whole. But just how large was the U.S. Navy in 1974? If one considers that back in 1974 our Navy active duty personnel number was daunting: nearly 546,000 sailors and officers, and if you add the civilian personnel and all of the support contractors it would easily top ONE MILLION people, one realizes that is in act a number in which a free-thinking individualistic citizen can easily drown. There is no simple answer as to why any 17-to-35 year old would make the commitment to sign-on through enlistment in order to serve our country on active duty, potentially risking life and limb in harm's way on a distant battlefield (*oceanic expanse*), but I will give it my best at describing the pros and cons of joining the vast ranks of the U.S. Armed Forces, and in particular the U.S. Navy.

For one, the tremendous resources that the U.S. Navy commands while plying the Oceans, Seas, and the vast totality of the Earth's surface and skies, literally able to reach into the most remote locations, simply cannot be matched, no matter whether we're talking about a global hospitality corporation, such as Hilton, or a global energy company like ExxonMobil, or even the aggregated military strengths of other nations. Here's the list of today's ten largest organizations:

Non-corporate employers are included in these lists:

Employer	Employees 2015	2010	Headquarters
United States Federal Government Includes DOD	3.2 million		United States
People's Liberation Army	2.3 million		China
Walmart	2.1 million		United States
McDonald's	1.9 million	1.7 million	United States
National Health Service	1.7 million	1.4 million	United Kingdom

Employer	Employees		Headquarters
	2015	2010	
China National Petroleum Corporation	1.6 million	1.7 million	China
State Grid Corporation of China	1.5 million	1.6 million	China
Indian Railways	1.4 million		India
Indian Armed Forces	1.3 million		India
Hon Hai Precision Industry (Foxconn)	1.3 million	0.8 million	Taiwan

These Navy resources include combat battle groups; land-sea-air-space weapons systems and the related command and control communications systems; a vast array of Training and Education centers, schools, and facilities; Medical Services that range from treatment of "bends" afflicting deep-sea divers, to post-natal care for dependent infants; a force-protection security framework; and a quality-of-life support organization. But how does the **individual** fit into, or better yet - what does the individual GET OUT OF this allegiance?

Individualism in the Navy

I'm no authority on the matter, but I can provide my perspective on the subject based on this six year slice of my life. Okay – generalizations usually don't hold water, but it's probably safe to say that we're all somewhat apprehensive about moving into a new school, a new job, a new neighborhood – it's the fear of the unknown, and the certainty that in the process of adapting to a new environment you'll necessarily have to make adjustments to your way of doing things. Your personal convictions might even have to be adjusted to conform to new norms.

For example, before enlisting in the Navy, I was dead certain that women did not belong on combatant ships, any combatant, whether submarine or aircraft carrier or all the various kinds of dreadnoughts in between. At the time I could not hypothetically picture having to rely on a woman to defend the ship by taking over the topside .50 caliber machine-gun that I had just been blown away from. My senses told me that as I lay writhing on the deck with shrapnel in my gut, Petty Officer Jane was more likely to attempt to contain my massive bleeding than she was to grab hold of the machine gun and strafe the advancing enemy. My conviction was that only Petty Officer Joe could effectively fill in for me since he would be much more likely to continue to rake the enemy with hot lead. Something about how a man's trained killer instinct can marry to that primordial survival instinct that we all have (again, generalizing!). To be part of a combatant ship's company, that was an essential quality, and in wartime to have even one member onboard without this quality was an unaffordable luxury.

My conviction had to give way in October 1978, in the wake of a court ruling that overturned statutes that forbade women from serving at sea. That's when the Navy launched the nascent *Women in Navy Ships* program and announced that they would assign 55 women officers and 375 female enlisted personnel to 21 ships during the next year. I suppose it's conceivable that the roles could easily be reversed, and that it would be Petty Officer Jane manning the .50 caliber gun from the onset, then getting blown away, and me coming up to the gun to replace her as she bled out onto the weather-deck. I would yell for a medic and in the same instant man the gun to keep raking our enemy with hot lead.

So I learned to adapt and adjust to new circumstances – one of the many essential components for change management.

Also consider the significance of the uniform, and how you are seen by people outside of the organization as just another number, another faceless "Government Issue – GI-Joe/GI-Jane" among the hundreds of thousands of other GIs who wear your same uniform. When you're on the inside, however, a sharp uniform becomes a source of pride. Earning awards builds confidence and self-esteem, so when you're on the inside, there's much more on the pro side about wearing your country's "suit of armor" so to speak. The military uniform is a window onto the person who wears it, and can denote pride, prowess, and personal achievement under duress, especially when the individual you're looking at shows battle awards and decorations. It's evident that we military persons instantly size-up another fellow in arms by the uniform he or she wears, and that applies whether we're facing a peer, a superior, or a subaltern. The uniform is in fact code for status, accomplishment, and personal potential. It's no wonder that Hollywood often glamorizes military uniforms as sexy and alluring.

From a practical perspective wearing a uniform sure made it easy for me to quickly and almost effortlessly be up and on the move early in the morning.

SIX Years Later

In about the same time that it took some of my High School friends to get their law degree and join a firm as an entry-level attorney, I had married a beautifully intelligent woman, fathered three lively sons, had seen much of the world –especially Europe-, had become proficient in Russian – Italian – and French while maintaining fluency in Spanish, had gone from the unassuming rank of E-1 to the fully-fledged professional technician rank of E-6, completed a Bachelor of Arts degree, and was on the verge of becoming a Commissioned Officer in the United States Navy through the OCS in Newport, Rhode Island. My instinct for constructive competition had been positively sharpened in my climb through the enlisted ranks of cryptologic technical expertise. This trait would remain one of my main strengths, well beyond retirement from military service, and I have to thank the Navy for imparting character under duress. The Navy had given context to that saying that my Dad has so often repeated to me when I was a young boy in elementary school: "When the going gets tough, the tough get going."

I now viewed the world through a transformed, worldlier maturity and embraced a more profound sense of responsibility for my actions and for the welfare and safety of my family and my shipmates. Over the course of those six years I had developed a profound sense of duty and obligation toward my Country, an obligation for which I was grateful. By forming a part of the massive combat and humanitarian relief organization that is the U.S. Navy, I had taken on a renewed sense of purpose, and my efforts toward personal and organizational improvement were now focused and deliberate. I realized that it was thanks to the multitude of officers, sailors, and support personnel who had preceded me that I was able to join ranks with an organization built on the distilled wisdom and achievements that had accrued over the course of 205 years, since the Navy's inception on October 13, 1775 by direction of the Second Continental Congress.

Now that OCS would be taking control of my life, and its chain of command would thrust me into yet another position of responsibility as a Battalion Commander to oversee the formative progress of well over 100 newly arrived college graduates and fellow "mustangs", I could honestly say that I not only knew what our "Non sibi sed patriae" motto stood for, but that I had in fact spent six years of my life representing the Navy's core values: Not for self, but for Country.

The fact that my wife Victoria and my three sons Ivan, Paul, and Daniel had also lived the intensely demanding Navy Family experience was something that I wore proudly, as a badge of honor. Their stalwart performance throughout all of the challenges that had come their way was simply awesome, and I was then, as I am now convinced that few families on the face of this Earth would be able to put up with the same adversities that they so ably shouldered. Sure, they had been subjected to many geo-moves: Monterey CA, San Angelo TX, Rota Spain, Cheltenham UK, and Newport RI – FIVE MOVES IN SIX YEARS!, but we collectively knew that not only had the family been hardened by the experience, but that each individual member of the family had learned many valuable lessons in life through the rough-and-tumble of Country above self.

As a Russian Interpretive Cryptologic Technician I also knew that my work had contributed directly to my Nation's progress toward winning the Cold War. While in 1980 we were still in the throes of the Arms Race, and much more competition over global hegemony was yet to be played out, I was proud to have contributed to the demystification of the Soviet Bear. This was not a Red Army of millions of ten-foot-tall soldiers with nuclear weapons that we were facing down. The Soviet Union represented another superpower on the world arena that was rife with vulnerabilities and limitations, and yet over the course of my six years as an observer of Soviet military operations, it was clear to me that the Soviets ran an organized military, and that the prospects of a global nuclear holocaust triggered by a first strike were indeed remote. The Soviets had built and imposed tight checks and balances on their military machine, and after the Cuban Missile Crisis had proved to the USA and our NATO allies that its control and command system was reliable, if not perfect. To ensure that was in fact the case, we, along with our NATO allies plied the skies, the oceans, and the seven continents with intelligence collection sensors that served a vast intelligence production engine designed to deliver clear Indications and Warning alerts to our strategic authorities. There was not a day, hour, or minute that this vast Western Powers intelligence network did not comb the environment with a fine tooth comb to detect any and all Soviet military transgressions against the stability of the post-WW-II world order.

From these six years of Service I was taking with me the confidence that my children and grandchildren would live to see a better day, unperturbed by the threat of Nuclear Holocaust. This was a containment strategy that I had served to help implement and maintain, and a strategy that in turn, when properly implemented, had given me the confidence to celebrate Liberty in the Land of the Free along with my family, my friends, and my military brothers and sisters, be they American, NATO Allies, or citizens of like-minded friendly nations from around the World.

IV. It's About What You've Given

There's a time-honored quotation from an historical speech given by President John F. Kennedy that has now become almost cliché it has been used and abused by so many people from so many nationalities and walks of life:

"Ask not what your Country can do for you, but what you can do for your Country."

In fact this phrase is ***almost*** as powerful as the U.S. Navy motto: "Not for self, but for Country." I say almost because the U.S. Navy motto packs the powerful notion that by adhering to that dictum, your very life is being invested into the success of your Country. There are many things that a citizen can do for the country, such as organize and lead a campaign for civil rights, deliver resources to an impoverished community, teach illiterate children to read and write, contribute to the strengthening of social justice, participate proactively in the democratic process. These are just examples of what a citizen can do for this great nation, yet what these actions all have in common is that they do not inherently put one's life in harm's way. To voluntarily step into the breach for the defense of the Nation requires a willingness to give oneself to a greater cause, knowing that when the winds of war are unleashed, your life is on the line.

Military service is a 24 hours-a-day, 7 days-a-week contract, unlike most other forms of human endeavor. Soldiers, Sailors, and Airmen are promised thirty days of vacation time per year, but all too often, especially in time of war, that freedom goes to the back-burner in order for servicemen and servicewomen to meet troop movement, mobilization, and deployment requirements. It's not uncommon to respond to unforeseen activity by potentially hostile military forces. Recent Cold War history is rich with examples of this type of response, examples such as the August 1978 Heightened Alert, when US Strategic

Air Command bombers, and US Navy anti-Submarine forces (across the entire range of air – surface – subsurface platforms) mobilized in response to Soviet submarine deployments in close proximity to the East coast of the USA.

Global events such as the Cyprus war, the civil war in Lebanon, the India-Pakistan war, the fall of the Shah of Iran in 1979, and the Soviet invasion of Afghanistan elicited massive and prolonged response from the U.S. Department of Defense, as Carrier Battle Groups, Army Brigades, Strategic Air Command wings became strategically positioned to react to whatever developments might put American lives in danger, and interfere with American security and economic interests. Often times our troops were spread thin in responding to such global conflicts, upsetting the established deployment cycles intended to provide service-members with adequate time at home.

The demand for force readiness is constant. That means that Sailors, Soldiers, and Airmen are routinely driven by training requirements that are tested in the crucible of military exercises large and small. These exercises are for the most part conducted with U.S. forces operating unilaterally. But larger scale exercises, such as those carried out by NATO on an annual basis under the banner of "Operation REFORGER" consist of large-scale force mobilizations that include major logistics coordination efforts by all NATO allies, with most of the burden falling on the USA, the UK, Canada and Germany. Given that REFORGER augmented NATO strengths in Germany, the VQ-2 Reconnaissance Squadron that I served routinely deployed to Germany to fly Baltic missions in an effort to collect on Soviet response to REFORGER. The exercise overlords in the European Command (EUCOM) Headquarters in Vaihingen Germany consistently valued the intelligence obtained from our reconnaissance operations, and for that very reason they routinely demanded more, which boosted morale tremendously, but also strained resources to the limit.

What often goes un-noticed is the leadership responsibility that senior personnel have when it comes to ensuring that junior personnel get the necessary training and military formation to carry out their demanding missions. The transfer of knowledge is one of the most important aspects of "giving", and I was constantly reminded of how my own personal cryptologic competencies had been forged by those who preceded me. By having personally experienced the significance of the "passing of the torch" when I was designated Airborne Cryptologic Crew-chief first for Skywarrior operations, and later for ARIES operations, I was prepared to in turn relay that torch of competency over to fellow cryptologists who were vying to take on the duties and responsibilities that I shouldered. The more precision, attention to detail, and thirst for knowledge that I could convey, the better my unit would be. My expectation was that if I could impart good momentum to my trainees that momentum would in turn gain in strength as they took on new junior personnel under their own wing. If their sense of purpose was even greater than mine, then the next wave of cryptologists would be that much better prepared to face new challenges. Such is the true spirit of camaraderie among the warrior class.

Today the phrase "Thank you for your Service" is almost an automatic refrain when a citizen becomes aware of a fellow citizen's military service. Seen from the outsider perspective, military service can at times represent an onerous obligation to bear arms for the good of the Country, yet at other times of peace and tranquility seem like an excursion that a fellow American has embarked on to get away from the doldrums of Joe six-pack's everyday struggle to make ends meet. Seen from the inside, however, servicemen and servicewomen are conscious of the fact that by wearing the uniform they do in fact represent the military power of the Nation. They, soldiers, airmen, sailors alike, are the very substance of that power behind the Seal of the President – the Commander in Chief. What they give in terms of time, intellect, and effort will become the bulwark of Democracy in times of war. Their personal lives and honor are willfully invested into forming part of a cohesive National Defense human entereprise.

To *Serve* is nothing more than to invest your persona into the Defense of the Nation for the common good, starting with your family, and expanding from the family nucleus outward. Eventually your ripple of good intentions will reflect and glimmer across the Sea of fellow citizens.

Epilogue

Duty, Honor, Country, these are the pillars of General Douglas McArthur's Thayer Award speech delivered on May 12th 1962 to the Cadets of the U.S. Military Academy at West Point[i]. His most recognizable words speak volumes of the American Warriors' ethos, and will endure beyond the life of our Republic, well into to a hypothetical time when a future generation may have to reach into ancient National Heritage to rebuild a worn code of conduct and renew a downtrodden military morale:

*"**Duty, honor, country: Those three hallowed words reverently dictate what you ought to be, what you can be, what you will be. They are your rallying point to build courage when courage seems to fail, to regain faith when there seems to be little cause for faith, to create hope when hope becomes forlorn.**"*

This bastion of moral courage was built by and belongs to the American Warrior, and to the society that supports and nourishes this powerful current of Patriotism. This is what propelled our Founders to defeat the British Empire in 1776, to survive the War of 1812 as a Nation with an identity that is today represented by the Anthem that Francis Scott Key penned as he witnessed the bombardment of Baltimore Maryland's Fort McHenry. This moral courage carried us through two World Wars, and underpinned our perseverance in winning the Cold War.

Yet in contrast to this boundless deep blue sea of national pride and military accomplishment there's the dissonance of treason. There are traitors who inherently lack the necessary courage to hold the line, and who from that condition of moral weakness abandon their comrades in arms because of cowardice. Traitors are also those whose undisclosed or undiscovered dementia results in horrendous "blue-on-blue / friendly fire" murder. But here I want to focus on yet another equally, if not more injurious form of treason. It's the one represented by those who commit espionage by delivering strategically sensitive information to America's existential enemies. In so doing, these spies endanger not only the lives of U.S. citizens, but they jeopardize and undermine whole-cloth operational campaigns. Espionage in time of war can endanger the very survival of the Nation.

Persons such as John Walker, his brother Arthur Walker, his friend Jerry Whitworth, and John's son Michael Walker (whose NCIS investigative report I would later be entrusted to study in order to implement countermeasures when I reported for duty onboard the USS NIMITZ in 1987) were all employed by the U.S. Navy. They got a paycheck, security clearances, and the trust of their colleagues as they literally spit on our Eagle while conducting treasonous spy-craft alongside their honorable duty-bound peers. These persons were categorically a breed apart, willing to trade the security and very lives of fellow sailors for a fistful of Kremlin-tainted dollars. Their minds and souls were fatally flawed by personal greed and treachery so that the very secrets that they had been entrusted to protect became the currency they used to betray their fellow Americans to line their pockets with blood money. John Walker brazenly sold top secret cryptographic keying materials that served to encrypt strategic nuclear submarine operations that held Soviet military arsenals in check, under the threat of immediate retaliation in case of all-out war. Even after he retired from the U.S. Navy, his spy network of family and friends continued to usurp national classified information to fatten their bank accounts. All of these grim characters remain behind bars for life, with the exception of Michael Walker who got off on parole by placing the blame on his father. To not shoulder responsibility for one's actions is anathema to the American Warrior ethos. Michael Walker walks free today, but his actions forever mark him as a failed citizen who on his own volition betrayed fellow servicemen.

More recently we've seen Private Bradley Manning (now Chelsea Manning, following the hormone treatment and gender-transition surgery that Manning underwent while incarcerated at Fort Leavenworth, Kansas) leak thousands of classified documents to WikiLeaks, the notoriously subversive Information Operations organization led by Julian Assange. Manning held the rank of "Specialist" intelligence analyst at the time of the leak activities, and was under oath to safeguard classified information under her care. Her crime of releasing sensitive classified information into the underground media, and from there into mainstream media where it was undoubtedly consumed by Al Qaida and others is only mitigated by the fact that the U.S. Army had been remiss in not having revoked Manning's access to classified information well before Manning reported for duty in Iraq. The telltale signs of Manning's deranged view of her place in the world had gone unchecked by her superiors over the course of several years before the leaks were committed. Manning may arguably have qualified as a conscientious objector in addition to being unfit for duty, in other words - definitely NOT the right person for assignment to intelligence operations in Iraq.

The final case of betrayal of Service that I will showcase as perfidious, and one that should be elevated to the pinnacle within the inferno reserved for American-born renegades is the case of Edward Snowden, a civilian contractor who singlehandedly orchestrated the largest violation of the Espionage Act in the history of the United States. Snowden had cut his teeth in the CIA as a trusted Cybersecurity expert, at one point serving as the Cybersecurity technical lead for the President's formal diplomatic visit to the 2008 NATO Summit in Romania. He separated from the CIA in 2009 to market his Top Secret – Sensitive Compartmented Information (TS/SCI) clearance and his familiarity with the CIA into the government contractor world by joining Dell, where, for reasons yet to be heard in a Federal trial of law, he soon began his "exfiltration" activities.

Snowden was an expert at concealing his frequent excursions into CIA, State Department, and Department of Defense on-line information repositories to download highly classified documents relating to U.S. intelligence operations around the globe. Over the course of his criminal information theft activities, he eventually took unlawful personal possession of over 200,000 sensitive, classified U.S. Government documents.

In his quest to reach into our Government's most sensitive archives he changed jobs from Dell where he had primarily supported the CIA's Information Technology (IT) needs, over to Booz Allen and Hamilton in support of the National Security Agency (NSA) IT requirements at its Hawaii facility in the spring of 2013. There was a concerted effort on Edward Snowden's part to amass important documents extracted not only from the CIA, the NSA, the DIA (Defense Intelligence Agency), but also from close allies such as the United Kingdom and Australia. In all there may have been close to one million sensitive documents and emails in his possession when he fled to Russia via Hong Kong in late 2013. While Snowden has attempted to justify his actions as a means to expose an "unconstitutional" form of espionage that NSA is supposedly conducting against the very own citizens that it serves, in reality the majority of the sensitive classified documents that Snowden flooded into the global media contain details relating to the conduct of U.S. – U.K. – Australian foreign intelligence operations. The damage that his activities have caused range from endangering counter-terrorist agents themselves to weakening US-UK-AUS counter-terrorism capabilities altogether by disclosing details of tactics and procedures used. These disclosures have helped terrorist and organized crime organizations to re-engineer their communications architectures, and to render useless many of the intelligence collection means employed by the governments of the U.S., the U.K., and Australia.

The dark cloud that Edward Snowden has cast over our concerted counter-terrorism operations signals the urgent need for his immediate extradition from Russia to the United States of America where, for the sake of this Republic and its stalwart NATO allies, Snowden must stand trial for his egregious transgressions of public trust. Only by facing the music can he take responsibility for his acts of treason.

We, those who Serve and live by the American Warrior Ethos, demand closure.

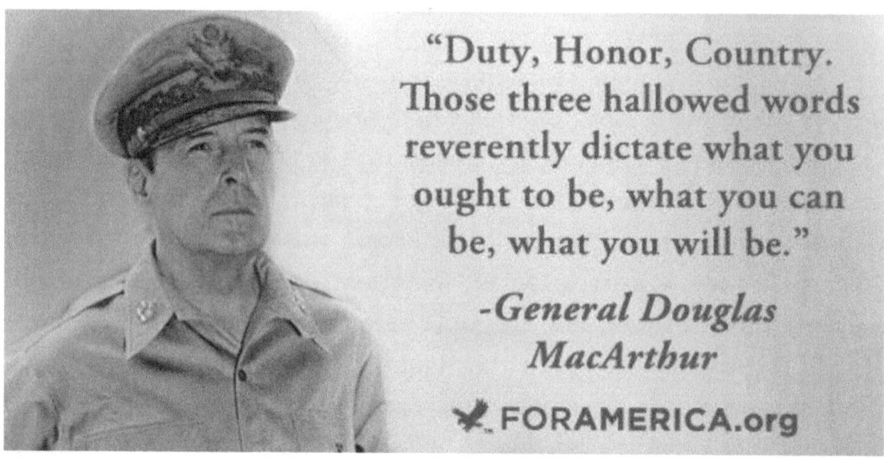

---

i http://www.jsums.edu/arotc/duty-honor-country/.